Undergraduate Texts in Mathematics

Undergraduate Texts in Mathematics

Undergraduate Texts in Mathematics are generally aimed at third- and fourth-year undergraduate mathematics students at North American universities. These texts strive to provide students and teachers with new perspectives and novel approaches. The books include motivation that guides the reader to an appreciation of interrelations among different aspects of the subject. They feature examples that illustrate key concepts as well as exercises that strengthen understanding.

For further volumes:
http://www.springer.com/series/666

John B. Conway

A Course in Point
Set Topology

 Springer

John B. Conway
Department of Mathematics
The George Washington University
Washington, DC, USA

ISSN 0172-6056   ISSN 2197-5604 (electronic)
ISBN 978-3-319-34834-6   ISBN 978-3-319-02368-7 (eBook)
DOI 10.1007/978-3-319-02368-7
Springer Cham Heidelberg New York Dordrecht London

Mathematics Subject Classification (2010): 54A05, 54E35, 54D20, 58C07

Printed on acid-free paper

Springer is part of Springer Science+Business Media (www.springer.com)

For Ann
How could I be so lucky!

Preface

Point set topology was my first love in mathematics. I took the course as an undergraduate at Loyola University in New Orleans and my professor, Harry Fledderman, told me to go to the library and solve all the problems in the book while he tutored the other student who had signed up for the course. (Yes, I know it sounds strange today, but there were only two students in the course.) I kept a notebook with my solutions, and once a week I reported for his inspection of my work. I felt like a real mathematician learning real mathematics. It had a great influence on me and made me realize how much I wanted to be a mathematician. Even now I can't tell you whether the love I have for point set topology was the cause of this feeling or whether that love was a consequence of this learning style. I was disappointed to later discover that research in this area had mostly petered out. I found equally attractive research areas in which to sow my oats, but I always retained this youthful love affair.

You can probably guess that I have long wanted to write a book on this topic, but other things took precedence. I am glad that was the case because now I think I have a better approach. I had an epiphany about halfway through my career when I realized I didn't have to teach my students everything I had learned about the subject at hand. I learned mathematics in school that I never used again, and not just because those things were in areas in which I never did research. At least part of this, I suspect, was because some of my teachers hadn't had this insight. Another reason is that many authors write textbooks as though they are writing a monograph directed at other faculty rather than thinking of the students as the audience. Also, mathematics refines and refreshes itself with time. Certain topics that were important at the inception of an area fade in significance, and some that are useful in various areas today must be added. Other topics are important, but only if you are one of that small percentage who specialize in a specific part of research; such things should not be taught to everyone who takes an introductory course. In addition, when a subject is developing, there is an emphasis on finding the intellectual boundaries of the concepts. Unless that viewpoint is abandoned when the subject is taught, it results in a greater prominence of pathology. An examination of early texts in any subject will reveal such an emphasis. With time, however, it is crucial to decide what should be taught in an introductory course such as the kind this book is written for. I see the purpose of a course in point set topology as giving the student a set of tools. The material is used in almost every part of mathematics.

In addition, I have come to believe that in teaching, it is best to go from the particular to the more general, an approach that has dominated my presentations in the last several texts I have written. To begin with, that is the way mathematics evolves. In addition, my experience is that all but the very best students find such an approach more digestible. The present book reflects my belief in this approach.

I see this text as aimed at an undergraduate audience that has had calculus and been exposed to the ideas of basic set theory like subsets, unions, intersections, functions, and little else. Nevertheless, I think it advisable that they have had at least a semester of analysis and been properly exposed to convergence and other topological notions in the real line. (I have included some appendices to help bridge the gap, but I am sure this will not suffice for all.) I also think point set topology is an excellent place to begin learning how to digest and write proofs. Thus, I tend to go slowly at the start of the book, including more detail than is needed for a seasoned student and even more than I include later in the text.

Following my philosophy of beginning with the particular, I start with metric spaces. I believe that these are far easier to connect with students' experience. They also seem to me to be the more prevalent topological spaces used in other areas and are therefore worth extra emphasis. Chapter 2 defines and develops abstract topological spaces, with metric spaces as the source of inspiration. I narrow the discussion by quickly restricting the focus to Hausdorff spaces. Needless to say, some of the more elementary arguments in topological spaces are the same as those in metric spaces. There is no problem here; I just refer students to the metric space proof and invite them to carry out the analogous argument, which in most cases is almost identical.

Chapter 3 concentrates on continuous real-valued functions. My belief is that the continuous functions on a space are more important than the underlying space. Maybe that's because I'm an analyst. I know that much of modern topology concentrates on the underlying geometry of a space, but surely that must be saved until after the student has encountered the need.

Biographies. In this book I continue the practice, started in a previous work, of including short biographical notes when a mathematician's result is mentioned. There is no scholarship on my part in this as all the material is from secondary sources, principally what I could find on the Web. In particular, I made heavy use of the site at the University of St Andrews `http://www-history.mcs.st-andrews.ac.uk/history/BiogIndex.html` and Wikipedia. I continue my practice of emphasizing personal aspects of these lives over their mathematical achievements, especially if there is something there that interests me.

Style. The longer I am a mathematician and the more I write, the more I ask myself questions about style. There is a difference between what you write and how you speak. That's true in mathematics just as it is outside of mathematics. I think I write more informally than many mathematicians, but there are some rules I try to follow even though they are not universal. One such rule is that with essentially no exceptions I use symbols like $\in$, $\subseteq$, $>$, and so on as verbs and only as verbs. (For example, $\in$ translates as "is an element of" and not as "in.") I think this consistency expedites reading. I experimented a long time ago with using such symbols only as prepositions but quickly decided this was awkward. When I am having a discussion in my office, I will frequently use them both ways, but when I write I try to stick with using them as verbs. So don't forget to read them that way.

Long ago I realized that every result is not a theorem. The label "Theorem" is reserved for the truly important results. It's not that those labeled "Proposition" are unimportant, but they may be more routine or, perhaps, they just don't have the impact on the development of the subject at hand. A corollary is a direct consequence of a proposition or a theorem. A lemma is a result whose usefulness is usually limited to the proof of the next result.

A Word to Students. If you want to learn mathematics, you cannot approach it as a spectator sport; sitting on the edge of the pool and dipping your toe in will not get you into the subject. You must jump into it with both feet, commit yourself, and do a lot of dirty work and splashing around before you can enter the profession at any level. In the course of this book there are many places where I leave proofs to readers as exercises; do them. When I give such an exercise, I think it is well within the scope of your ability provided you understood the concepts. Doing those exercises will confirm that you understand what has come before; if you cannot do them, it may mean you overlooked something and should go back.

Throughout the text you will see words like *Verify!* and *Why?* I am trying to put a speed bump in your reading. I want you to be sure you understand what was just said.

Sometimes my exercises ask a question. A basic part of mathematics is deciding whether something is true and then proving it. Mathematicians are constantly trying to discover whether something is true and are seldom, if ever, presented with a known truth and asked to prove it. So when you see such an exercise and you think something is true, you must prove it; if you think it false, you must find a counterexample. (The ability to manufacture examples is a precious talent that you should cultivate.)

For most of you this is the class where you will perfect your ability to execute a proof. A proof is a written explanation of why a particular statement is true. If you're a veteran at writing proofs, ignore the rest of this paragraph. For the less experienced, here is some advice about discovering and writing down a proof. (a) Write down the hypothesis under the label GIVEN. (b) Write down the conclusion under the label TO SHOW. (c) Be sure you understand all the terms used in (a) and (b). (d) If you have difficulty, try drawing a picture. Look at an example. Try rephrasing the hypothesis and conclusion. Try to construct a counterexample. (Sometimes this makes you see why you cannot get a counterexample and, hence, why the proposition is true.) (e) If all this fails, be extremely clever.

Have fun, live long, and prosper.

Washington, DC   John B. Conway

Contents

Chapter 1

Metric Spaces

We want to abstract the idea of distance because this notion often arises in mathematics. The concept must be general enough to encompass the circumstances under which it arises, but it should conform to the intuitive notion we all have of what is meant by the distance between two points. Because this is being done at the start of the first section, before proceeding it would be profitable for readers to reflect on what properties they think should be included in an abstract concept of distance; then they can compare these thoughts with the definition given below.

1.1. Definitions and Examples

Here is the accepted mathematical definition of an abstract distance.

Definition 1.1.1. A *metric space* is a pair (X, d) where X is a set and d is a function $d : X \times X \to [0, \infty)$, called a *metric*, that satisfies the following properties for all x, y, z in X:

(a) $d(x, y) = d(y, x)$;

(b) $d(x, y) = 0$ if and only if $x = y$;

(c) (the triangle inequality) $d(x, y) \leq d(x, z) + d(z, y)$.

These three properties are usually part of what we intuitively would associate with the idea of distance. Condition (a) is sometimes called the reflexive property and says that the distance from x to y is the same as the distance from y to x. The second property says the obvious: the distance from a point to itself is 0 and the only point at a distance zero from x is x itself. The third, the triangle property, says that the shortest distance between two points is the direct one—not a distance involving a third point. One might be tempted to phrase this by saying that the shortest distance between two points is a straight line, but in this abstract setting we have no concept of a line, let alone a straight one.

J.B. Conway, *A Course in Point Set Topology*, Undergraduate Texts in Mathematics, DOI 10.1007/978-3-319-02368-7_1,
© Springer International Publishing Switzerland 2014

It is certainly possible that some readers might have additional properties they would want a distance function to have, but the aforementioned
properties are the minimal ones. Indeed, in mathematics there are several
particular situations where additional axioms for a distance are assumed;
those are more specialized theories, and what we are now going to explore is
this basic one. Here are some examples of metric spaces.

Example 1.1.2. (a) Let $X = \mathbb{R}$, the set of real numbers, and define $d(x, y)$
$= |x - y|$. See Exercise 1.

(b) Let $X = \mathbb{R}^2$, the plane, and define $d((x_1, y_1), (x_2, y_2)) = [(x_1 - x_2)^2 +$
$(y_1 - y_2)^2]^{\frac{1}{2}}$. Readers know from the Pythagorean Theorem that this is
the straight-line distance, and they can use geometry to verify that this
standard notion of the distance between two points satisfies the axioms
in the preceding definition. More generally, we can define a metric on
q-dimensional Euclidian space $\mathbb{R}^q$ by

$$d(x, y) = \left[\sum_{n=1}^{q} (x_n - y_n)^2 \right]^{\frac{1}{2}}$$

for $x = (x_1, \ldots, x_q)$ and $y = (y_1, \ldots, y_q)$ in $\mathbb{R}^q$. However, proving that
this satisfies the needed axioms, specifically the triangle inequality, requires some effort, and the proof will be given later (Corollary 1.1.5).

(c) Let $X = \mathbb{R}^q$, q-dimensional Euclidean space, for x, y in $\mathbb{R}^q$ define

$$d(x, y) = \sum_{n=1}^{q} |x_n - y_n|.$$

This is easier to verify as a metric (Exercise 2).

(d) Again let $X = \mathbb{R}^q$, and now define

$$d(x, y) = \max\{|x_n - y_n| : 1 \le n \le q\}.$$

Once again, $(\mathbb{R}^q, d)$ is a metric space (Exercise 3). It is worth observing
that in each of the last three examples, when $q = 1$, all these metrics are
the standard absolute value on $\mathbb{R}$.

(e) Let X be any set, and define

$$d(x, y) = \begin{cases} 0 & \text{if } x = y, \\ 1 & \text{if } x \neq y. \end{cases}$$

It is a simple exercise to verify that (X, d) is a metric space. This is called
the *discrete metric* on X.

(f) An important class of examples arises as follows. Suppose (X, d) is given. If Y is a nonempty subset of X, then (Y, d) is a metric space and is referred to as a *subspace*. As a specific instance of this we can take $X = \mathbb{R}$ and $Y = [0, 1]$.

Now let us show that the function d given in part (b) of the preceding example is indeed a metric. To do this we need a famous inequality. Before presenting this inequality, we introduce the helpful notation that for vectors $x = (x_1, \ldots, x_q)$ and $y = (y_1, \ldots, y_q)$ in $\mathbb{R}^q$, $\langle x, y \rangle = \sum_{n=1}^{q} x_n y_n$. Actually, this is more than just "helpful" notation as it denotes the inner or "dot" product in the vector space $\mathbb{R}^q$. This connection will not be explored here, however, and we will only regard this as notation. It is useful to observe the following properties for all vectors x, y, z in $\mathbb{R}^q$ and all real numbers t.

1.1.3
$$\begin{cases} \langle x, x \rangle \geq 0, \\ \langle x, y \rangle = \langle y, x \rangle, \\ \langle tx + z, y \rangle = t\langle x, y \rangle + \langle z, y \rangle, \\ \langle x, y + tz \rangle = \langle x, y \rangle + t\langle x, z \rangle. \end{cases}$$

Theorem 1.1.4 (Cauchy[1]–Schwarz[2] inequality). *If* $x = (x_1, \ldots, x_q)$ *and* $y = (y_1, \ldots, y_q)$ *are vectors in* $\mathbb{R}^q$, *then*

$$\left[\sum_{n=1}^{q} x_n y_n \right]^2 \leq \left[\sum_{n=1}^{q} x_n^2 \right] \left[\sum_{n=1}^{q} y_n^2 \right].$$

Proof. First note that if we use the inner product notation introduced previously, then the sought-after inequality becomes

$$\langle x, y \rangle^2 \leq \langle x, x \rangle \langle y, y \rangle.$$

---

[1]Augustin Louis Cauchy was born in Paris in August 1789, a month after the storming of the Bastille. He was educated in engineering, and his first job was in 1810 working on the port facilities at Cherbourg in preparation for Napoleon's contemplated invasion of England. In 1812 he returned to Paris, and his energies shifted toward mathematics. His contributions were monumental, with a plethora of results bearing his name. His collected works fill 27 published volumes. As a human being he left much to be desired. He was highly religious with a totally dogmatic personality, often treating others with dismissive rudeness. Two famous examples were his treatment of Abel and Galois: he refused to consider their highly significant works, which they had submitted to him. Both Abel and Galois died young. Perhaps better treatment by Cauchy would have given them some recognition that would have resulted in a longer life and a productive career to the betterment of mathematics; we'll never know. He had two doctoral students, one of whom was Bunyakowsky. Cauchy died in 1857 in Sceaux near Paris.

[2]Hermann Amandus Schwarz was a German mathematician born in 1843 in Hermsdorf, Silesia, now part of Poland. He began his studies at Berlin in chemistry but switched to mathematics and received his doctorate in 1864 under the direction of Weierstrass. He held positions at Halle, Zurich, Göttingen, and Berlin. His work centered on various geometry problems that were deeply connected to analysis. This included work on surfaces and conformal mappings in analytic function theory, any student of which will see his name in prominence. He died in Berlin in 1921.

Using (1.1.3) we have

$$
\begin{aligned}
0 &\le \langle x - ty, x - ty \rangle \\
&= \langle x, x \rangle - t\langle y, x \rangle - t\langle x, y \rangle + t^2 \langle y, y \rangle \\
&= \langle x, x \rangle - 2t\langle x, y \rangle + t^2 \langle y, y \rangle \\
&= \gamma - 2\beta t + \alpha t^2 \equiv q(t),
\end{aligned}
$$

where $\gamma = \langle x, x \rangle, \beta = \langle x, y \rangle, \alpha = \langle y, y \rangle$. Thus $q(t)$ is a quadratic polynomial in the variable t. Since $q(t) \ge 0$ for all t, the graph of $q(t)$ stays above the x-axis, except that it might be tangent at a single point; that is, $q(t) = 0$ has at most one real root. From the quadratic formula we get that $0 \ge 4\beta^2 - 4\alpha\gamma = 4(\beta^2 - \alpha\gamma)$. Therefore,

$$
0 \ge \beta^2 - \alpha\gamma = \langle x, y \rangle^2 - \langle x, x \rangle \langle y, y \rangle,
$$

proving the inequality.   ∎

Corollary 1.1.5. *If* $d : \mathbb{R}^q \times \mathbb{R}^q \to [0, \infty)$ *is defined as in* Example 1.1.2(b), *then* d *is a metric.*

Proof. We begin by noting that $d(x, y) = \sqrt{\langle x - y, x - y \rangle}$. Using the Cauchy–Schwarz inequality (1.1.3) and the vector space properties of $\mathbb{R}^q$ we get

$$
\begin{aligned}
d(x, y)^2 &= \langle x - y, x - y \rangle \\
&= \langle (x - z) + (z - y), (x - z) + (z - y) \rangle \\
&= \langle x - z, x - z \rangle + 2\langle x - z, z - y \rangle + \langle z - y, z - y \rangle \\
&\le d(x, z)^2 + 2\sqrt{\langle x - z, x - z \rangle}\sqrt{\langle z - y, z - y \rangle} + d(z, y)^2 \\
&= d(x, z)^2 + 2d(x, z)d(z, y) + d(z, y)^2 \\
&= [d(x, z) + d(z, y)]^2 .
\end{aligned}
$$

Taking square roots shows that the triangle inequality holds. The remainder of the proof that d defines a metric is straightforward. (Verify!)   ∎

When $x \in X$ and $r > 0$, we introduce the notation

$$
B(x; r) = \{y \in X : d(x, y) < r\}, \quad \overline{B}(x; r) = \{y \in X : d(x, y) \le r\}.
$$

The set $B(x; r)$ is called the *open ball* about x, or centered at x, of radius r; $\overline{B}(x; r)$ is called the *closed ball* about x of radius r. If $X = \mathbb{R}$, then $B(x; r)$ is the open interval $(x - r, x + r)$ and $\overline{B}(x; r)$ is the closed interval $[x - r, x + r]$. If $X = \mathbb{R}^2$, then $B(x; r)$ is the so-called "open" ball or disk centered at x of radius r that does not include the bounding circle, and $\overline{B}(x; r)$ is the corresponding "closed" disk that does include the bounding circle. The use of the words and here will be made clear momentarily. Meanwhile, notice that when $s < r$, $\overline{B}(x; s) \subseteq B(x; r)$. This trivial observation will come in handy.

Definition 1.1.6. If (X, d) is a metric space, then a subset G of X is *open* if for each x in G there is an $r > 0$ such that $B(x; r) \subseteq G$. A subset F of X is *closed* if its complement, $X \backslash F$, is open.

Example 1.1.7. (a) We observe that X and $\emptyset$ are both open and closed sets. In fact, it is clear that X is open and $\emptyset$ is open since the condition is vacuously satisfied. That is, since there are no points in $\emptyset$, every point in $\emptyset$ satisfies the condition needed for $\emptyset$ to be an open set. Since both these sets are open, their complements are closed.

(b) For any $r > 0$, $B(x; r)$ is open. If $y \in B(x; r)$ and $0 < s < r - d(x, y)$, then $B(y; s) \subseteq B(x; r)$. In fact, if $d(y, z) < s$, then $d(x, z) \le d(x, y) + d(y, z) < d(x, y) + s < r$. Thus $B(y; s) \subseteq B(x; r)$.

(c) For any $r > 0$, $\overline{B}(x; r)$ is closed. To see this, set $G = X \backslash B(x; r)$ and let $y \in G$; thus, $d(y, x) > r$. Let $0 < s < d(x, y) - r$. If $d(z, y) < s$, then $r < d(x, y) - s < d(x, y) - d(y, z) \le [d(x, z) + d(z, y)] - d(y, z) = d(x, z)$; that is, $B(y; s) \subseteq G$. Since this shows that G is open, it follows that $\overline{B}(x; r)$ is closed.

(d) Any finite subset of X is closed. In fact, if $F = \{x_1, \ldots, x_n\}$ and $x \in X \backslash F$, then we can find a positive radius $r < \min\{d(x, x_1), \ldots, d(x, x_n)\}$ and $B(x; r) \subseteq X \backslash F$.

It is important when discussing open and closed sets to be conscious of the universe. When (X, d) is a metric space and $Y \subseteq X$, we have that (Y, d) is also a metric space [Example 1.1.2(f)]. To say that we have an open set A in (Y, d) does not mean that A is open in X. Note that in such a circumstance when $y \in Y$ and $r > 0$, the open ball about y of radius r is $B_Y(y; r) = \{z \in Y : d(z, y) < r\} = B(y; r) \cap Y$. This may not be an open set in X. For example, if $X = \mathbb{R}$ and $Y = [0, 1]$, then $[0, \frac{1}{2})$ is open as a subset of Y but not as a subset of X. Another example: $B_Y(\frac{1}{4}; \frac{1}{3}) = [0, 7/12)$. When we want to emphasize the open and closed sets in the subspace metric (Y, d), we will use the terms *open relative to Y* or *relatively open* in Y and *closed relative to Y* or *relatively closed in Y*. The proof of the next proposition is Exercise 5.

Proposition 1.1.8. *Let (X, d) be a metric space, and let Y be a subset of X.*

(a) *A subset G of Y is relatively open in Y if and only if there is an open subset U in X with $G = U \cap Y$.*

(b) *A subset F of Y is relatively closed in Y if and only if there is a closed subset D in X such that $F = D \cap Y$.*

Before getting into the properties of open and closed sets, let us remark that in establishing part (c) of Example 1.1.7 we essentially derived a useful inequality that we now complete. It is often called the *reverse triangle inequality*.

Proposition 1.1.9. *If (X, d) is a metric space and $x, y, z \in X$, then*

$$|d(x, y) - d(y, z)| \le d(x, z).$$

Proof. In part (c) of Example 1.1.7 we showed that as a consequence of the triangle inequality, $d(x, y) - d(y, z) \leq d(x, z)$. Now reverse the roles of x and z in this inequality and we get that $d(z, y) - d(y, x) \leq d(z, x)$, from which we also have that $d(y, z) - d(x, y) \leq d(x, z)$.   ∎

Now we return to an examination of open and closed sets. For the remainder of the chapter (X, d) will be a metric space that is always on our radar screen, even though it may not be explicitly mentioned.

Proposition 1.1.10. (a) *If* $G_1, \ldots, G_n$ *are open sets, then* $\bigcap_{k=1}^{n} G_k$ *is open.*
(b) *If* $\{G_i : i \in I\}$ *is a collection of open sets, then* $\bigcup_{i \in I} G_i$ *is open.*

Proof. (a) Let $x \in \bigcap_{k=1}^{n} G_k$. So for $1 \leq k \leq n$ there is an $r_k > 0$ with $B(x, r_k) \subseteq G_k$. If $r = \min\{r_1, \ldots, r_n\}$, then $r > 0$ and $B(x; r) \subseteq \bigcap_{k=1}^{n} G_k$.
(b) The proof of this is even easier than the proof of part (a). If $x \in \bigcup_{i \in I} G_i$, then for some j in I, $x \in G_j$; so there is an open ball $B(x; r)$ contained in G_j. But then $B(x; r) \subseteq \bigcup_{i \in I} G_i$.   ∎

By taking complements and using De Morgan's laws (Proposition A.1.5), we get the following proposition for closed sets.

Proposition 1.1.11. (a) *If* $F_1, \ldots, F_n$ *are closed sets, then* $\bigcup_{k=1}^{n} F_k$ *is closed.*
(b) *If* $\{F_i : i \in I\}$ *is a collection of closed sets, then* $\bigcap_{i \in I} F_i$ *is closed.*

We will see that the open and closed sets in a metric space play a central role. The next result is a prelude to this.

Definition 1.1.12. Let A be a subset of X. The *interior* of A, denoted by int A, is the set defined by

$$\text{int } A = \bigcup \{G : G \text{ is open and } G \subseteq A\}.$$

The *closure* of A, denoted by cl A, is the set defined by

$$\text{cl } A = \bigcap \{F : F \text{ is a closed and } A \subseteq F\}.$$

The *boundary* of A, denoted by ∂A, is the set defined by

$$\partial A = \text{cl } A \cap \text{cl } (X \backslash A).$$

Let us note that there is always an open set contained in any set A—namely, the empty set, $\emptyset$. It may be, however, that $\emptyset$ is the only open set contained in A, in which case int $A = \emptyset$. Similarly, X is a closed set containing any set A; but it may be the only such set, in which case cl $A = X$. (We will have more to say about this latter case subsequently.) It follows from the preceding two propositions that int A is open (though possibly empty) and cl A is closed (though possibly equal to X). We have that int $\emptyset = \emptyset = \text{cl } \emptyset$ and int $X = X = \text{cl } X$. Before looking at more meaningful examples, it would be profitable to first prove some properties of the closure and interior of a set.

Proposition 1.1.13. *Let $A \subseteq X$.*

(a) $x \in \operatorname{int} A$ *if and only if there is an $r > 0$ such that $B(x;r) \subseteq A$.*

(b) $x \in \operatorname{cl} A$ *if and only if for every $r > 0$, $B(x;r) \cap A \neq \emptyset$.*

Proof. (a) If $B(x;r) \subseteq A$, then since $B(x;r)$ is open, we have that $B(x;r)$ $\subseteq \operatorname{int} A$; hence $x \in \operatorname{int} A$. Now assume that $x \in \operatorname{int} A$. So there is an open set G such that $x \in G \subseteq A$. But since G is open, there is a radius $r > 0$ with $B(x;r) \subseteq G$, and we have established the converse.

(b) Suppose $x \in \operatorname{cl} A$. If $r > 0$, then $B(x;r)$ is open and $X \backslash B(x;r)$ is closed. It cannot be that $A \subseteq X \backslash B(x;r)$ since, by definition, this implies $\operatorname{cl} A \subseteq X \backslash B(x;r)$, contradicting the fact that $x \in \operatorname{cl} A$. Thus $B(x;r) \cap A \neq \emptyset$. Now assume that $x \notin \operatorname{cl} A$; that is, $x \in X \backslash \operatorname{cl} A$, an open set. By definition there is a radius $r > 0$ such that $B(x;r) \subseteq X \backslash \operatorname{cl} A$. So for this radius, $B(x;r) \cap A = \emptyset$. ∎

The preceding proposition is very useful as it provides a concrete, one-point-at-a-time method to determine the closure and the interior of a set. We will see this in the following example.

Example 1.1.14. (a) Sometimes things can become weird with interiors and closures. Consider the metric space $\mathbb{R}$ and the subset $\mathbb{Q}$ of all rational numbers. We are assuming the reader is familiar with the fact that if a and b are two real numbers with $a < b$, then there is a rational number x with $a < x < b$ (Axiom A.3.2). Note that this says that $\operatorname{cl} \mathbb{Q} = \mathbb{R}$. In fact, if $x \in \mathbb{R}$, then $B(x;r) = (x - r, x + r)$, and this interval must contain a rational number. By the preceding proposition, $x \in \operatorname{cl} \mathbb{Q}$. We also have that $\operatorname{int} \mathbb{Q} = \emptyset$. To see this again use the preceding proposition and the fact that between any two real numbers there is an irrational number(Axiom A.3.2). This means that when $x \in \mathbb{Q}$, no open ball $B(x;r)$ can be contained in $\mathbb{Q}$ so that $\operatorname{int} \mathbb{Q} = \emptyset$. Using the same reasoning we see that $\operatorname{cl}[\mathbb{R} \backslash \mathbb{Q}] = \mathbb{R}$ and $\operatorname{int}[\mathbb{R} \backslash \mathbb{Q}] = \emptyset$. (Verify!) It follows that $\partial \mathbb{Q} = \mathbb{R}$.

(b) Here is a cautionary tale. Since $\overline{B}(x;r)$ is closed, we have that $\operatorname{cl} B(x;r) \subseteq \overline{B}(x;r)$. It may not be, however, that $\operatorname{cl} B(x;r) = \overline{B}(x;r)$. In fact, suppose that $X = \{(0,0)\} \cup \{(a,b) : a^2 + b^2 = 1\} \subseteq \mathbb{R}^2$. So X consists of the origin in the plane together with the unit circle centered at the origin. Give X the metric it inherits as a subset of $\mathbb{R}^2$. In this case, $\{(0,0)\} = \operatorname{cl} B((0,0);1) \neq \overline{B}((0,0);1) = X$. It is also true that when X is a discrete metric space as defined in Example 1.1.2(e), then $B(x;1) = \operatorname{cl} B(x;1)$, whereas $\overline{B}(x;1) = X$.

The next proposition contains some useful information about closures and interiors of sets. Its proof is left as Exercise 7.

Proposition 1.1.15. *Let A be a subset of X.*

(a) A *is closed if and only if* $A = \operatorname{cl} A$.

(b) A *is open if and only if* $A = \operatorname{int} A$.

(c) $\operatorname{cl} A = X \backslash [\operatorname{int}(X \backslash A)]$, $\operatorname{int} A = X \backslash \operatorname{cl}(X \backslash A)$, *and* $\partial A = \operatorname{cl} A \backslash \operatorname{int} A$.

(d) *If* $A_1, \ldots, A_n$ *are subsets of* X, *then* $\operatorname{cl} \left[\bigcup_{k=1}^n A_k \right] = \bigcup_{k=1}^n \operatorname{cl} A_k$.

Part (d) of the preceding proposition does not hold for the interior. For example, if $X = \mathbb{R}$, $a < b < c$, $A = (a, b]$, and $B = [b, c)$, then $\operatorname{int}(A \cup B) = (a, c)$, while $\operatorname{int} A \cup \operatorname{int} B = (a, b) \cup (b, c)$. Also see Exercises 8 and 9.

Finally, we want to examine how we put together metric spaces to obtain new metric spaces. We will see later (Theorem 2.6.6) how to combine a sequence of metric spaces to obtain a new one, but now let us concentrate on putting together a finite number. We will focus on showing how to combine just two spaces because the process for a finite number of spaces is exactly the same but with more complicated notation. Recall that the cartesian product of two sets X_1 and X_2 is defined as $X_1 \times X_2 = \{(x_1, x_2) : x_1 \in X_1, x_2 \in X_2\}$.

Definition 1.1.16. If (X_1, d_1) and (X_2, d_2) are two metric spaces, then define the new metric space $(X_1 \times X_2, d)$ by letting

$$d((x_1, x_2), (y_1, y_2)) = d_1(x_1, y_1) + d_2(x_2, y_2)$$

for all x_1, y_1 in X_1 and x_2, y_2 in X_2.

Verifying that this does define a metric is Exercise 10. We could define the metric in other ways, for example,

$$d((x_1, x_2), (y_1, y_2)) = \max\{d_1(x_1, y_1), d_2(x_2, y_2)\}.$$

We will see later that these and other appropriate definitions yield "equivalent metrics." But that must be postponed until we have a bit more background. Right now we will always use the metric given previously in Definition 1.1.16.

Proposition 1.1.17. *Adopt the previously given notation.*

(a) *If* G_1, G_2 *are open sets in* X_1, X_2, *then* $G_1 \times G_2$ *is open in* $X_1 \times X_2$.

(b) *If* F_1, F_2 *are closed sets in* X_1, X_2, *then* $F_1 \times F_2$ *is closed in* $X_1 \times X_2$.

(c) *If* G *is open in* $X_1 \times X_2$ *and* $(x_1, x_2) \in G$, *then there is an* $r > 0$ *such that* $B(x_1; r) \times B(x_2; r) \subseteq G$.

Proof. (a) If $(x_1, x_2) \in G_1 \times G_2$, then for $k = 1, 2$ let $r_k > 0$ such that $B(x_k; r_k) \subseteq G_k$. It follows that if $r = \min\{r_1, r_2\}$, then $B((x_1, x_2); r) \subseteq G_1 \times G_2$.

(b) $(X_1 \times X_2) \backslash (F_1 \times F_2) = [(X_1 \backslash F_1) \times X_2] \cup [X_1 \times (X_2 \backslash F_2)]$, and by (a) this is the union of two open sets.

(c) Let $\epsilon > 0$ such that $B((x_1, x_2); \epsilon) \subseteq G$. If $0 < r < \epsilon/2$, then this works. ∎

We close this section with a topic we will encounter in several places as we proceed through the book.

Definition 1.1.18. A subset E of a metric space (X, d) is *dense* if cl $E = X$. A metric space (X, d) is *separable* if it has a countable dense subset.

We made reference to this concept just after defining the closure of a set. So a set E is dense if and only if X is the only closed subset of X that contains E.

Example 1.1.19. (a) Every metric space is dense in itself.
  (b) The rational numbers form a dense subset of $\mathbb{R}$, as do the irrational numbers. This is a rephrasing of Example 1.1.14(a). This might also explain the name given to Axiom A.3.2. We note that this implies that $\mathbb{R}$ is separable since $\mathbb{Q}$ is countable (Corollary A.5.5).
  (c) The set of all points in $\mathbb{R}^q$ with rational coordinates is dense in $\mathbb{R}^q$. This follows from the preceding example, and it also says that $\mathbb{R}^q$ is separable by Proposition A.5.4. Also see Exercise 13.
  (d) If X is any set and d is a discrete metric on X [Example 1.1.2(e)], then the only dense subset of X is X itself. In fact, if E is a dense subset of (X, d) and $x \in X$, then it must be that $B(x; 1/2) \cap E \neq \emptyset$; but from the definition of the discrete metric it follows that $B(x; 1/2) = \{x\}$.

In part (d) of the preceding example we used the next result, and we record it here for future reference.

Proposition 1.1.20. *A set E is dense in (X, d) if and only if for every x in X and every $r > 0$, $B(x; r) \cap E \neq \emptyset$.*

The proof is an easy application of Proposition 1.1.13.

Exercises

(1) Verify the statement in Example 1.1.2(a).
(2) Verify the statement in Example 1.1.2(c).
(3) Verify the statement in Example 1.1.2(d).
(4) In the Cauchy–Schwarz inequality, show that equality holds if and only if the vectors x and y are linearly dependent.
(5) Prove Proposition 1.1.8. [Hint: if G is a relatively open subset of Y, then for each y in G let $r_y > 0$ such that $B_Y(y; r_y) \subseteq G$. Now consider $\bigcup \{B(y; r_y) : y \in G\}$.]
(6) If Y is a subset of X, then consider the metric space (Y, d) and suppose $Z \subseteq Y$. (a) Show that H is a relatively open subset of Z if and only if there is a relatively open subset H_1 of Y such that $H = H_1 \cap Z$. (b) Show that D is a relatively closed subset of Z if and only if there is a relatively closed subset D_1 of Y such that $D = D_1 \cap Z$. (Hint: use Proposition 1.1.8.)
(7) Prove Proposition 1.1.15.

(8) Show that if $A_1, \ldots, A_n$ are subsets of X, then int $[\bigcap_{k=1}^{n} A_k] = \bigcap_{k=1}^{n}$ int A_k.

(9) Show that Proposition 1.1.15(d) does not hold for infinite unions and the preceding exercise does not hold for infinite intersections.

(10) Verify that the function d given in Definition 1.1.16 is a metric.

(11) Define $\rho : X_1 \times X_2 \to [0, \infty)$ by

$$\rho((x_1, x_2), (y_1, y_2)) = \max\{d_1(x_1, y_1), d_2(x_2, y_2)\}.$$

(a) Show that ρ is a metric on $X_1 \times X_2$. (b) Show that a set is open in $(X_1 \times X_2, d)$ if and only if it is open in $(X_1 \times X_2, \rho)$.

(12) Let ℓ^∞ denote the set of all bounded sequences of real numbers; that is, ℓ^∞ consists of all sequences $\{x_n\}$ such that $x_n \in \mathbb{R}$ for all $n \geq 1$ and $\sup_n |x_n| < \infty$. If $x = \{x_n\}, y = \{y_n\} \in \ell^\infty$, then define $d(x, y) = \sup_n |x_n - y_n|$. (a) Show that d defines a metric on ℓ^∞. (b) If e_n denotes a sequence with a 1 in the nth place and zeroes elsewhere, show that $B(e_n; \frac{1}{2}) \cap B(e_m; \frac{1}{2}) = \emptyset$ when $n \neq m$. (c) Is the set $\{e_n : n \geq 1\}$ closed?

(13) If for $k = 1, 2$, A_k is a dense subset of the metric space (X_k, d_k), then show that $A_1 \times A_2$ is a dense subset of $X_1 \times X_2$. Hence the product of a finite number of separable spaces is separable.

1.2. Sequences and Completeness

Recall that throughout the chapter (X, d) is a given metric space that is under consideration. Here we will discuss sequences as an extension of the same concept encountered in calculus. Recall that a sequence is just a way of enumerating some points in the space X: $x_1, x_2, \ldots$; this is denoted by $\{x_n\}$. Precisely or technically this is a function from the natural numbers $\mathbb{N}$ into X: $n \mapsto x_n$. By the way, we will sometimes change the domain of this function to $\mathbb{N} \cup \{0\}$ so that we get a sequence $\{x_0, x_1, \ldots\}$; or maybe it might be changed to get $\{x_2, x_3, \ldots\}$. There is no essential difference; we have a specific beginning and a countably infinite following. So the set of all integers, $\mathbb{Z}$, is not permitted as an indexing set.

Definition 1.2.1. A sequence $\{x_n\}$ in X *converges* to x if for every $\epsilon > 0$ there is an integer N such that $d(x, x_n) < \epsilon$ when $n \geq N$. The notation for this is $x_n \to x$ or $x = \lim_n x_n$.

We note that when $X = \mathbb{R}$, this is exactly the definition of a convergent sequence learned in calculus. We also emphasize that the integer N in the definition depends on ϵ. Generally, the smaller the value of ϵ, the larger we must make N (Exercise 1). We might also mention that the inequality $d(x, x_n) < \epsilon$ can easily be replaced by $d(x_n, x) \leq \epsilon$. Why?

Example 1.2.2. (a) If (X, d) is a discrete metric space, then a sequence $\{x_n\}$ in X converges to x if and only if there is an integer N such that $x_n = x$ whenever $n \geq N$.

(b) If (X, d) is the cartesian product of the two metric spaces (X_1, d_1) and (X_2, d_2), then a sequence $\{(x_n^1, x_n^2)\}$ in X converges to (x^1, x^2) if and only if $x_n^1 \to x^1$ and $x_n^2 \to x^2$.

The value of sequences and the concept of convergence begins to surface in the next proposition.

Proposition 1.2.3. *A subset F of X is closed if and only if whenever $\{x_n\}$ is a sequence in F and $x_n \to x$, it follows that $x \in F$.*

Proof. First assume that F is closed, $\{x_n\}$ is a sequence of elements in F, and $x_n \to x$. If it were the case that $x \notin F$, then the fact $X \backslash F$ is open would mean there is an $r > 0$ such that $B(x; r) \subseteq X \backslash F$. But then there would be an N such that for $n \geq N$, $d(x_n, x) < r$, that is, $x_n \in B(x; r) \subseteq X \backslash F$, a contradiction. Hence it must be that $x \in F$. Now assume the sequential condition is satisfied. If $x \in \operatorname{cl} F$, then Proposition 1.1.13 implies that $B(x; r) \cap F \neq \emptyset$ for every $r > 0$. In particular, for every natural number n there is a point x_n in $B(x; n^{-1}) \cap F$. Hence $\{x_n\}$ is a sequence in F and $d(x_n, x) < n^{-1}$; thus $x_n \to x$, and so $x \in F$. That is, $\operatorname{cl} F \subseteq F$, and so F is closed. ∎

Definition 1.2.4. If $A \subseteq X$, then a point x in X is called a *limit point* of A if for every $\epsilon > 0$ there is a point a in $B(x; \epsilon) \cap A$ with $a \neq x$.

The emphasis here is that no matter how small we take ϵ, we can find such a point a different from x that belongs to $B(x; \epsilon) \cap A$. It is not required that x must belong to A for it to be a limit point (more on that later). If x is not a limit point of A and, in addition, belongs to A, then it is called an *isolated point* of A.

Example 1.2.5. (a) Let $X = \mathbb{R}$ and $A = (0, 1) \cup \{2\}$. Every point in $[0, 1]$ is a limit point of A, but 2 is not. In fact, 2 is an example of an isolated point of A, the only isolated point of A.

(b) If $X = \mathbb{R}$ and $A = \mathbb{Q}$, then every point of X is a limit point of A and A has no isolated points.

(c) If $X = \mathbb{R}$ and $A = \{n^{-1} : n \in \mathbb{N}\}$, then 0 is a limit point of A, while the points n^{-1} are all isolated points.

We will also need the idea of a *subsequence* as in calculus. So if $\{x_n\}$ is a sequence in X, then a subsequence is another sequence $\{x_{n_1}, x_{n_2}, \dots\}$, denoted by $\{x_{n_k}\}$, where $n_1 < n_2 < \cdots$. Technically, this is the function $k \mapsto x_{n_k}$. We leave the proof of the following proposition as Exercise 6.

Proposition 1.2.6. *If $x_n \to x$ in X and $\{x_{n_k}\}$ is a subsequence, then $x_{n_k} \to x$.*

Proposition 1.2.7. *Let A be a subset of the metric space X.*

(a) *A point x is a limit point of A if and only if there is a sequence of distinct points in A that converges to x.*

(b) *A is a closed set if and only if it contains all its limit points.*

(c) cl $A = A \cup \{x : x$ is a limit point of $A\}$.

Proof. (a) Suppose $\{a_n\}$ is a sequence of distinct points in A such that $a_n \to x$. If $\epsilon > 0$, then there is an N such that $a_n \in B(x; \epsilon)$ for $n \geq N$. Since the points in $\{a_n\}$ are distinct, there is at least one point different from x. Thus x is a limit point. Now assume that x is a limit point. Let $a_1 \in A \cap B(x; 1)$ such that $a_1 \neq x$. Let $\epsilon_2 = \min\{2^{-1}, d(x, a_1)\}$; so there is a point $a_2 \in A \cap B(x; \epsilon_2)$ with $a_2 \neq x$. Note that $a_2 \neq a_1$.

Claim. There is a sequence of positive numbers $\{\epsilon_n\}$ and a sequence of distinct points $\{a_n\}$ in A such that: (i) $\epsilon_n \leq n^{-1}$; (ii) $a_n \neq x$ for all $n \geq 1$; (iii) $d(x, a_n) < \epsilon_n$.

We establish this claim by induction. By taking $\epsilon_1 = 1$, we have that the claim holds when $n = 1$. We also showed that the claim holds when $n = 2$, and that indicates how to carry out the induction step. Here are the details. Assume the claim holds for some integer n, and let $\epsilon_{n+1} = \min\{(n + 1)^{-1}, d(x, a_1), \ldots, d(x, a_n)\}$. So there is a point a_{n+1} in $A \cap B(x; \epsilon_{n+1})$ with $a_{n+1} \neq x$. Note that parts (i), (ii), and (iii) are satisfied with $n + 1$ in place of n. Also, $a_{n+1} \neq a_k$ for $1 \leq k \leq n$ since $d(x, a_{n+1}) < \epsilon_{n+1} \leq d(x, a_k)$ for $1 \leq k \leq n$. By induction we have that the claim is valid; by the claim we have a sequence of distinct points in A that converges to x.

(b) *and* (c). Clearly (b) will follow once we prove (c). Let B denote the set on the right-hand side of the equation in (c). By Proposition 1.2.3 and part (a) we have that $B \subseteq$ cl A. On the other hand, if $x \in$ cl A, then Proposition 1.2.3 implies there is a sequence $\{a_n\}$ in A such that $a_n \to x$. Either $\{a_n\}$ has an infinite number of distinct terms or a finite number. In the first case there is a subsequence $\{a_{n_k}\}$ of distinct terms; by (a), x is a limit point. Thus $x \in A$. In the second case there is a subsequence $\{a_{n_k}\}$ that is constant; thus $a_{n_k} = x$ for all $k \geq 1$, and so $x \in A$. ∎

Definition 1.2.8. If $A \subseteq X$ and $x \in X$, then the *distance* from x to A is

$$\mathrm{dist}\,(x, A) = \inf\{d(x, a) : a \in A\}.$$

Clearly, when $x \in A$, dist $(x, A) = 0$. But it is possible for the distance from a point to a set to be 0 when the point is not in the set, as we now see.

Proposition 1.2.9. *If $A \subseteq X$, then* cl $A = \{x \in X : \mathrm{dist}\,(x, A) = 0\}$.

Proof. If $x \in$ cl A, then there is a sequence $\{a_n\}$ in A such that $a_n \to x$; so dist $(x, a_n) \to 0$, and it follows that dist $(x, A) = 0$. Conversely, if dist $(x, A) = 0$, then there is a sequence $\{a_n\}$ in A such that $d(x, a_n) \to 0$. Thus, $a_n \to x$, and so $x \in$ cl A. ∎

Also see Exercise 4.

Definition 1.2.10. A sequence $\{x_n\}$ in X is a *Cauchy sequence* if for every $\epsilon > 0$ there is an integer N such that $d(x_n, x_m) < \epsilon$ whenever $m, n \geq N$. The metric space X is said to be *complete* if every Cauchy sequence converges.

It is rather easy to see that in any metric space, every convergent sequence is a Cauchy sequence. In fact, if $x_n \to x$ and $\epsilon > 0$, then choose N such that $d(x, x_n) < \epsilon/2$. Hence, when $m, n \geq N$, $d(x_n, x_m) \leq d(x_n, x) + d(x, x_m) < \epsilon$. Therefore, complete metric spaces are those for which convergent and Cauchy sequences are the same. The virtue of the concept of being Cauchy, when the space is complete, is that you know the limit exists without having to produce the point x that is the limit.

Before we see some examples, let us establish some facts about Cauchy sequences.

Proposition 1.2.11. *If $\{x_n\}$ is a Cauchy sequence and some subsequence of $\{x_n\}$ converges to x, then $x_n \to x$.*

Proof. Suppose $x_{n_k} \to x$, and let $\epsilon > 0$. Choose an integer N_1 such that $d(x_{n_k}, x) < \epsilon/2$ for $n_k \geq N_1$, and choose an integer N_2 such that $d(x_n, x_m) < \epsilon/2$ when $m, n \geq N_2$. Set $N = \max\{N_1, N_2\}$, and let $n \geq N$. Fix any $n_k \geq N$. Since we have that $n_k \geq N_1$ and both n and n_k are larger than N_2, we get that $d(x, x_n) \leq d(x, x_{n_k}) + d(x_{n_k}, x_n) < \epsilon/2 + \epsilon/2 = \epsilon$. ∎

For any set E define its *diameter* as

$$\operatorname{diam} E = \sup\{d(x, y) : x, y \in E\}$$

It is easy to see that $\operatorname{diam} E = \operatorname{diam}[\operatorname{cl} E]$.

Theorem 1.2.12 (Cantor's[3] Theorem). *A metric space (X, d) is complete if and only if whenever $\{F_n\}$ is a sequence of nonempty subsets satisfying* (i) *each F_n is closed;* (ii) *$F_1 \supseteq F_2 \supseteq \cdots$;* (iii) *$\operatorname{diam} F_n \to 0$, then $\bigcap_{n=1}^{\infty} F_n$ is a single point.*

Proof. Assume (X, d) is complete and $\{F_n\}$ is as in the statement of the theorem. For each n let $x_n \in F_n$. If $\epsilon > 0$, then let N be such that $\operatorname{diam} F_n < \epsilon$ for $n \geq N$. Thus, if $m, n \geq N$, then (ii) implies $x_n, x_m \in F_N$ and so $d(x_n, x_m) \leq \operatorname{diam} F_N < \epsilon$. Thus $\{x_n\}$ is a Cauchy sequence; since (X, d) is complete, there is an x in X such that $x_n \to x$. Since each F_n is closed, $x \in \bigcap_{n=1}^{\infty} F_n$. If there is another point y in $\bigcap_{n=1}^{\infty} F_n$, then $d(x, y) \leq \operatorname{diam} F_n$ for each $n \geq 1$. By (iii), $y = x$.

Now assume that (X, d) satisfies the stated conditions and $\{x_n\}$ is a Cauchy sequence. Set $F_n = \operatorname{cl}\{x_n, x_{n+1}, \dots\}$. Clearly, (i) and (ii) are satisfied. If $\epsilon > 0$, then let N be such that $d(x_n, x_m) < \epsilon$ for $m, n \geq N$. But for $k \geq N$, $\operatorname{diam} F_k = \sup\{d(x_n, x_m) : m, n \geq k\} \leq \epsilon$. Thus $\{F_n\}$ satisfies the

---

[3]Georg Cantor was the child of an international family. His father was born in Denmark and his mother was Russian; he himself was born in 1845 in St. Petersburg, where his father was a successful merchant and stockbroker. Cantor is recognized as the father of set theory, having invented cardinal and ordinal numbers and proved that the irrational numbers are uncountable. He received his doctorate from the University of Berlin in 1867 and spent most of his career at the University of Halle. His work was a watershed event in mathematics, but it was condemned by many prominent contemporary mathematicians. The work was simply too radical, with counterintuitive results such as $\mathbb{R}$ and $\mathbb{R}^q$ having the same number of points. He began to suffer from depression around 1884. This progressed and plagued him the rest of his life. He died in a sanatorium in Halle in 1918.

three conditions, so that $\bigcap_{n=1}^{\infty} F_n = \{x\}$ for some point x. But for any $n \geq 1$, $d(x, x_n) \leq \operatorname{diam} F_n \to 0$. Therefore, $x_n \to x$, and (X, d) is complete. $\blacksquare$

Example 1.2.13. (a) The first example of a complete metric space is $\mathbb{R}$. This is a consequence of various properties of the real number system seen in § A.3 as follows. Let $\{F_n\}$ be a sequence of nonempty subsets of $\mathbb{R}$ satisfying the three conditions in Cantor's Theorem. Since each has finite diameter, they must be bounded. By the Completeness Property (Axiom A.3.3) of $\mathbb{R}$, $a_n = \inf F_n$ and $b_n = \sup F_n$ exist. Since F_n is closed, Corollary A.3.5 implies that $a_n, b_n \in F_n$ and so $0 \leq b_n - a_n \leq \operatorname{diam} F_n \to 0$. Since $F_{n+1} \subseteq F_n$, it follows that $a_n \leq a_{n+1} \leq b_{n+1} \leq b_n$. By Proposition A.3.6 there are points a and b such that $a_n \to a$ and $b_n \to b$. But it must be that $a, b \in F_n$ for each n and so $|b - a| \leq \operatorname{diam} F_n \to 0$; so $a = b$. That is, $\bigcap_{n=1}^{\infty} F_n$ is the singleton $\{a\}$. By Cantor's Theorem, $\mathbb{R}$ is complete.

(b) For any $d \geq 1$, $\mathbb{R}^q$ is complete. In fact if $x_n = (x_n^1, \dots, x_n^q)$ and $\{x_n\}$ is a Cauchy sequence in $\mathbb{R}^q$, then $|x_n^k - x_m^k| \leq d(x_n, x_m)$ for $1 \leq k \leq q$ and so it follows that each $\{x_n^k\}$ is a Cauchy sequence in $\mathbb{R}$. By (a), $x_n^k \to x^k$ for some real number x^k. It follows that $x_n \to x = (x^1, \dots, x^q)$. (Verify!)

The proof of the next proposition is Exercise 8.

Proposition 1.2.14. *If (X, d) is a complete metric space and $Y \subseteq X$, then (Y, d) is complete if and only if Y is closed in X.*

It is not hard to find examples of metric spaces that are not complete. For example using the preceding proposition we can look at any subset of $\mathbb{R}$ that is not closed. $\mathbb{Q}$ is one such and a dramatic one at that.

We close this section by dwelling a bit on the concept of the diameter of a set, introduced just prior to the statement of Cantor's Theorem.

Definition 1.2.15. Say that a subset A of (X, d) is *bounded* if $\operatorname{diam} A < \infty$.

Proposition 1.2.16. (a) *A subset A of (X, d) is bounded if and only if for any x in X there is an $r > 0$ such that $A \subseteq B(x; r)$.*

(b) *The union of a finite number of bounded sets is bounded.*

(c) *A Cauchy sequence in (X, d) is a bounded set.*

Proof. (a) If $A \subseteq B(x; r)$, then $\operatorname{diam} A \leq 2r$, so that A is bounded. Conversely, assume that A is bounded with δ as its finite diameter. Fix a point x in X and some point a_0 in A. For any point a in A, $d(x, a) \leq d(x, a_0) + d(a_0, a) \leq d(x, a_0) + \delta$. If we let $r = 2[d(a_0, x) + \delta]$, then $A \subseteq B(x; r)$.

(b) If A_k is bounded for $1 \leq k \leq n$ and $x \in X$, then let $r_k > 0$ such that $A_k \subseteq B(x; r_k)$. If we set $r = r_1 + \cdots + r_n$, then $A_1 \cup \cdots \cup A_n \subseteq B(x; r)$.

(c) If $\{x_n\}$ is a Cauchy sequence, then there is an $N \geq 1$ such that $d(x_n, x_m) < 1$ for $m, n \geq N$. If $B = \{x_n : n \geq N\}$, this says that diam $B \leq 1$, so that B is bounded. On the other hand, $A = \{x_1, \ldots, x_N\}$ is bounded since finite sets are bounded. By part (b), $\{x_n\} = A \cup B$ is bounded.

■

Exercises

(1) For each of the following sequences $\{x_n\}$ find the value of the limit, and for each stipulated value of ϵ find a value of N such that $d(x, x_n) < \epsilon$ when $n \geq N$. (a) $X = \mathbb{R}, x_n = n^{-1}, \epsilon = 0.0001$. (b) $X = \mathbb{R}, x_n = e^{-n}, \epsilon = 0.0001$. (c) $X = \mathbb{R}^{q+1}, x_n = (n^{-1}, n^{-2}, \ldots, n^{-q}, e^{-n}), \epsilon = 0.0001$. (Are the values for N you found the smallest possible? This has no bearing on the convergence, but it is a bit more challenging to find the smallest possible N.)

(2) Verify the statements made in Example 1.2.2.

(3) Suppose $\{x_n\}$ is a sequence in X that converges to x and $z_1, \ldots, z_m$ is a finite collection of points in X. Define a new sequence $\{y_n\}$ in X by letting $y_k = z_k$ for $1 \leq k \leq m$ and $y_k = x_{k-m}$ when $k \geq m+1$. Show that $y_n \to x$.

(4) If $A \subseteq X$, show that int $A = \{x : \text{dist}\,(x, X \backslash A) > 0\}$. Can you give an analogous characterization of ∂A?

(5) (a) If $A \subseteq X$, show that $x \in \text{cl}\,A$ if and only if x is either a limit point of A or an isolated point of A. (b) Show that if a set has no limit points, it is closed. (c) Give an example of an infinite subset of $\mathbb{R}$ that has no limit points.

(6) Prove Proposition 1.2.6.

(7) For the real line, consider the three restrictions (i), (ii), and (iii) placed on the sets $\{F_n\}$ in Cantor's Theorem. (a) Find a sequence of sets $\{F_n\}$ that satisfies (i) and (ii), but $\bigcap_{n=1}^{\infty} F_n = \emptyset$. (b) Find a sequence of sets $\{F_n\}$ that satisfies (i) and (iii), but $\bigcap_{n=1}^{\infty} F_n = \emptyset$. (c) Find a sequence of sets $\{F_n\}$ that satisfies (ii) and (iii), but $\bigcap_{n=1}^{\infty} F_n = \emptyset$. (d) Show that if $\{F_n\}$ is a sequence of bounded sets satisfying (i) and (ii), then $\bigcap_{n=1}^{\infty} F_n \neq \emptyset$, and give an example where it is not a singleton.

(8) Prove Proposition 1.2.14.

(9) Show that a sequence $\{(x_1^n, x_2^n)\}$ in $\mathbb{R}^q \times \mathbb{R}^m$ converges to (x_1, x_2) if and only if the same thing happens when we consider the sequence as belonging to $\mathbb{R}^{q+m}$.

(10) Let (X, d) be the cartesian product of the two metric spaces (X_1, d_1) and (X_2, d_2). (a) Show that a sequence $\{(x_n^1, x_n^2)\}$ in X is a Cauchy sequence in X if and only if $\{x_n^1\}$ is a Cauchy sequence in X_1 and $\{x_n^2\}$ is a Cauchy sequence in X_2. (b) Show that X is complete if and only if both X_1 and X_2 are complete.

(11) Show that the metric space ℓ^∞ defined in Exercise 1.1.12 is complete.

1.3. Continuity

Here we will extend the concept of a continuous function seen in calculus to a mapping between two metric spaces.

Definition 1.3.1. If (X, d) and (Z, ρ) are two metric spaces, a function $f : X \to Z$ is *continuous at a point a* in X if for every $\epsilon > 0$ there is a $\delta > 0$ such that when $d(a, x) < \delta$, it follows that $\rho(f(a), f(x)) < \epsilon$. f is said to be a *continuous function* if it is continuous at each point of X.

Note that if $X = Z = \mathbb{R}$, then this becomes the statement that for every $\epsilon > 0$ there is a $\delta > 0$ such that when $|x - a| < \delta$, we have that $|f(a) - f(x)| < \epsilon$, the precise definition from calculus. The next result should have a familiar ring if we put it in this calculus setting.

Proposition 1.3.2. *If (X, d) and (Z, ρ) are metric spaces and $f : X \to Z$, then f is continuous at a if and only if whenever $\{x_n\}$ is a sequence in X and $x_n \to a$, then $f(x_n) \to f(a)$ in Z.*

Proof. Suppose f is continuous at a and $x_n \to a$ in X. If $\epsilon > 0$, then let $\delta > 0$ such that when $d(a, x) < \delta$, it follows that $\rho(f(a), f(x)) < \epsilon$. Let $N \geq 1$ such that $d(x_n, a) < \delta$ when $n \geq N$. Thus, $\rho(f(x_n), f(a)) < \epsilon$ when $n \geq N$. Since ϵ was arbitrary, this says that $f(x_n) \to f(a)$. To prove the converse, assume that f is not continuous at a. So there exists an $\epsilon > 0$ such that for every $\delta > 0$ there is at least one x with $d(x, a) < \delta$, but $\rho(f(x), f(a)) \geq \epsilon$. In particular, taking $\delta = n^{-1}$ we have that for every $n \geq 1$ there is an x_n with $d(x_n, a) < n^{-1}$ and $\rho(f(x_n), f(a)) \geq \epsilon$. But this says that $x_n \to a$, and $\{f(x_n)\}$ does not converge to $f(a)$. ∎

We will not spend much time investigating functions continuous at a single point, but we will have much to say about functions continuous on the entire metric space.

Theorem 1.3.3. *If (X, d) and (Z, ρ) are metric spaces and $f : X \to Z$, then the following statements are equivalent.*

(a) *f is a continuous function on X.*

(b) *If U is an open subset of Z, then $f^{-1}(U)$ is an open subset of X.*

(c) *If D is a closed subset of Z, then $f^{-1}(D)$ is a closed subset of X.*

Proof. (b) *is equivalent to* (c). Note that

$$f^{-1}(Z \backslash U) = X \backslash f^{-1}(U) \text{ and } f^{-1}(Z \backslash D) = X \backslash f^{-1}(D).$$

From these equalities the equivalence of the two statements is straightforward.

(a) *implies* (b). Let $a \in f^{-1}(U)$ so that $\alpha = f(a) \in U$. Since U is open, there is an $\epsilon > 0$ such that $B(\alpha; \epsilon) \subseteq U$. Since f is continuous there is a $\delta > 0$ such that $d(a, x) < \delta$ implies $\rho(f(a), f(x)) < \epsilon$. In other words, $B(a; \delta) \subseteq f^{-1}(B(\alpha; \epsilon)) \subseteq f^{-1}(U)$. Since a was an arbitrary point in $f^{-1}(U)$, this says that $f^{-1}(U)$ is open.

(b) *implies* (a). If $a \in X$ and $\epsilon > 0$, then $B(f(a); \epsilon)$ is open; so by (b) we have that $f^{-1}(B(f(a); \epsilon))$ is an open set in X that contains a. Thus there is a $\delta > 0$ such that $B(a; \delta) \subseteq f^{-1}(B(f(a); \epsilon)$. That is, $d(a, x) < \delta$ implies $\rho(f(a), f(x)) < \epsilon$, and so f is continuous at a. ∎

Now to generate some examples of continuous functions. We will assume that the reader is already familiar with the continuity of the various functions encountered in calculus, such as the trigonometric functions, the exponential, and the logarithm, as well as various functions defined on subsets of $\mathbb{R}^q$. The first result is in this general area of functions defined on $\mathbb{R}^q$. (See Exercise 1.2.9.)

Example 1.3.4. (a) The function from $\mathbb{R}^q \times \mathbb{R}^q \to \mathbb{R}^q$ defined by $(x, y) \mapsto x + y$ (vector addition) is continuous. Verifying this is easy if we use Proposition 1.3.2 and Exercise 1.2.10.
(b) The function from $\mathbb{R} \times \mathbb{R}^q \to \mathbb{R}^q$ defined by $(t, x) \mapsto tx$ (scalar multiplication) is continuous.
(c) If (X, d) is any metric space, then $x_0 \in X$, and we define $f : X \to \mathbb{R}$ by $f(x) = d(x, x_0)$, then f is continuous. In fact, the reverse triangle inequality (Proposition 1.1.9) says that $|f(x) - f(y)| \leq d(x, y)$, from which continuity follows from either the definition or Proposition 1.3.2.
(d) If (X, d) is a discrete metric space, then the only continuous functions from $[0, 1]$ into (X, d) are the constant functions.

Recall the definition of the distance from a point to a set A, dist (x, A), given in the last section.

Proposition 1.3.5. *If (X, d) is a metric space and $A \subseteq X$, then*
$$|\text{dist}\,(x, A) - \text{dist}\,(y, A)| \leq d(x, y)$$
for all x, y in X.

Proof. If $a \in A$, then $d(x, a) \leq d(x, y) + d(y, a)$; thus, taking the infimum over all a in A we get dist $(x, A) \leq \inf\{d(x, y) + d(y, a) : a \in A\} = d(x, y) +$ dist (y, A). Reversing the roles of x and y we have dist $(y, A) \leq d(x, y) +$ dist (x, A), whence we get the inequality. ∎

Corollary 1.3.6. *If A is a nonempty subset of X, then $f : X :\to \mathbb{R}$ defined by $f(x) = \text{dist}\,(x, A)$ is a continuous function.*

We also have the following result, possibly expected by the reader, whose proof follows easily by using Proposition 1.3.2.

Proposition 1.3.7. *If (X, d) is a metric space and f and g are continuous functions from X into $\mathbb{R}$, then $f + g : X \to \mathbb{R}$ and $fg : X \to \mathbb{R}$ are continuous, where $(f + g)(x) = f(x) + g(x)$ and $(fg)(x) = f(x)g(x)$ for all x in X. If $f(x) \neq 0$ for all x in X, then $f^{-1} = 1/f : X \to \mathbb{R}$ defined by $f^{-1}(x) = 1/f(x) = [f(x)]^{-1}$ is a continuous function.*

The last proposition is a way to combine continuous functions to obtain another continuous function. Here is another. Recall that if $f : X \to Z$ and $g : Z \to W$ are functions, then the *composition* of f and g is the function $g \circ f : X \to W$ defined by $g \circ f(x) = g(f(x))$.

Proposition 1.3.8. *The composition of two continuous functions is also continuous.*

Proof. If $f : X \to Z$ and $g : Z \to W$, then for any subset G of W, then we have that $(g \circ f)^{-1}(G) = f^{-1}[g^{-1}(G)]$. Thus, if G is an open subset of W, it follows that $(g \circ f)^{-1}(G)$ is open in X. Hence $g \circ f$ is continuous. ∎

Later in the book we will present several results on manufacturing continuous functions from a metric space into the real numbers. We have seen examples of such continuous functions on specific metric spaces, but we want to show the existence of continuous functions with specific properties on arbitrary ones. Here we use Corollary 1.3.6 to prove a famous result that we will prove later in a more general context.

Theorem 1.3.9 (Urysohn's[4] Lemma). *If A and B are two disjoint closed subsets of X, then there is a continuous function $f : X \to \mathbb{R}$ having the following properties:*

(a) $0 \le f(x) \le 1$ *for all x in X:*

(b) $f(x) = 0$ *for all x in A:*

(c) $f(x) = 1$ *for all x in B.*

Proof. Define $f : X \to \mathbb{R}$ by

$$f(x) = \frac{\operatorname{dist}(x, A)}{\operatorname{dist}(x, A) + \operatorname{dist}(x, B)},$$

which is well defined since the denominator never vanishes. (Why?) It is easy to check that f has the desired properties. ∎

Corollary 1.3.10. *If F is a closed subset of X and G is an open set containing F, then there is a continuous function $f : X \to \mathbb{R}$ such that $0 \le f(x) \le 1$ for all x in X, $f(x) = 1$ when $x \in F$, and $f(x) = 0$ when $x \notin G$.*

Proof. In Urysohn's Lemma, take A to be the complement of G and $B = F$. ∎

---

[4]Pavel Samuilovich Urysohn was born in 1898 in Odessa, Ukraine. He was awarded his habilitation in June 1921 from the University of Moscow, where he remained as an instructor. He began his work in analysis but switched to topology, in which he made several important contributions, especially in developing a theory of dimension. His work attracted attention from the mathematicians of the day, and in 1924 he set out for a tour of the major universities in Germany, Holland, and France, meeting with Hausdorff, Hilbert, and others. That same year, while swimming off the coast of Brittany, France, he drowned. He is buried in Batz-sur-Mer in Brittany. In just 3 years he left his mark on mathematics.

Recall that a map $f : X \to Z$ is *injective* if it is one-to-one; that is, if $f(x) = f(y)$, then $x = y$. The function is *surjective* if it is onto; that is, for any z in Z there is a point x in X with $f(x) = z$. If f is both injective and surjective, then it is said to be *bijective*. When f is bijective, we can define the function $f^{-1} : Z \to X$ by letting $f^{-1}(z)$ equal the unique point x in X such that $f(x) = z$. The reader may have noticed the possibility of confusing notation with this definition of f^{-1} and that in Proposition 1.3.7. I am afraid this is something we will have to live with. Usually, the context will make it clear which definition we are talking about. We are coming into contact with mathematical tradition or custom, and we will not fight it. If there is ever the possibility of confusion, we can always stick with the notation $1/f$ to denote the reciprocal of a function.

Definition 1.3.11. If (X, d) and (Z, ρ) are metric spaces, then a map $f : X \to Z$ is called a *homeomorphism* if f is bijective and both f and f^{-1} are continuous. Two metric spaces are said to be *homeomorphic* if there is a homeomorphism from one onto the other.

Note that a bijection $f : X \to Z$ is a homeomorphism precisely when a sequence $\{x_n\}$ in X converges to x if and only if $f(x_n) \to f(x)$. A homeomorphism identifies the two spaces. That is, via f they have the same convergent sequences, open sets, closed sets, and continuous functions. In fact, homeomorphisms define an equivalence relation (Definition 2.8.1) between metric spaces. To be sure, it is not the only possible equivalence relation, and perhaps the idea of an isometry, a map $f : (X, d) \to (Z, \rho)$ satisfying $\rho(f(x), f(y)) = d(x, y)$ for all x, y in X, seems more natural. Indeed, isometries identify the two structures, and it readily follows that an isometry is a homeomorphism. Nevertheless, we will emphasize homeomorphisms rather than isometries. While an isometry identifies the *metric* structures of two metric spaces, a homeomorphism identifies the two *topological* structures. That is, it identifies the open sets, closed sets, etc. This is more in keeping with the subject of this book.

Definition 1.3.12. If X is a set, then the two metrics d and ρ are said to be *equivalent* if they define the same convergent sequences. Equivalently, d and ρ are equivalent if the identity map $i : (X, d) \to (X, \rho)$ is a homeomorphism.

Exercise 1.1.11 shows that the metric ρ defined there is equivalent to the usual metric.

Proposition 1.3.13. *For any metric space* (X, d)

$$\rho(x, y) = \frac{d(x, y)}{1 + d(x, y)}$$

defines an equivalent metric.

Proof. The first step is to show that ρ is a metric, and the only point of contention is whether it satisfies the triangle inequality. This we do by an examination of the function $f(t) = t/(1+t)$ defined on $(-1, \infty)$. Note that for $t > -1$, $f'(t) = (1+t)^{-2} > 0$ and $f''(t) = -2(1+t)^{-3} < 0$. Thus, for $s, t > 0$ and $g(t) = f(s) + f(t) - f(s+t)$, $g'(t) = f'(t) - f'(s+t) > 0$. So $g(t) > g(0) = 0$. This says that $f(s+t) \leq f(s) + f(t)$ for all s, t in $[0, \infty)$. Since f is increasing, for all x, y, z in X,

$$
\begin{aligned}
\rho(x, y) &= \frac{d(x, y)}{1 + d(x, y)} \\
&\leq \frac{d(x, z) + d(z, y)}{1 + d(x, z) + d(z, y)} \\
&= f[d(x, z) + d(z, y)] \\
&\leq \frac{d(x, z)}{1 + d(x, z)} + \frac{d(z, y)}{1 + d(z, y)} \\
&= \rho(x, z) + \rho(z, y),
\end{aligned}
$$

so that ρ satisfies the triangle inequality.

Now to show that the two metrics are equivalent. Clearly, if $d(x_n, x) \to 0$, then $\rho(x_n, x) = d(x_n, x)[1 + d(x_n, x)]^{-1} \to 0$. Conversely, if $\rho(x_n, x) \to 0$, then, since we always have that $\rho(x, y) < 1$, $d(x_n, x) = \rho(x_n, x)[1 - \rho(x_n, x)]^{-1} \to 0$. ∎

An important feature of the preceding metric is that it is bounded by 1. This is a warning for us not to put too much stock in the concept of a bounded set.

Example 1.3.14. Two equivalent metrics do not necessarily have the same Cauchy sequences; in fact, with one metric it can be complete and with respect to the other it is not. We will show this by defining an equivalent metric ρ on $\mathbb{R}$ such that with this metric $(\mathbb{R}, \rho)$ is not complete. Consider the circle $X = \{(x, y) \in \mathbb{R}^2 : x^2 + (y-1)^2 = 1\}$, that is, the circle of radius one in the plane centered at $(0, 1)$. Give X the metric d it has as a subset of the plane. We want to describe a function $f : \mathbb{R} \to X$ geometrically. It is possible to do this with a formula, but the geometry makes all we are going to say transparent. For any t in $\mathbb{R}$ consider the straight line in $\mathbb{R}^2$ determined by $(t, 0)$ and $(0, 2)$, and let $f(t)$ equal the point on the circle X where this line intersects it. Note several things. If $-1 < t < 1$, then $f(t)$ lies on the lower half of the circle, whereas when $|t| > 1$, $f(t)$ is on the upper half; also, $f(1) = (1, 0)$, $f(-1) = (-1, 0)$. In addition, note that f is injective and $f(\mathbb{R}) = X \backslash \{(0, 2)\} \equiv Y$. We use f to put a new metric on $\mathbb{R}$ by letting $\rho(t, s) = d(f(t), f(s))$. It is not hard to see that the metric ρ is equivalent to the standard metric on the real line defined by the absolute value and that $f : (\mathbb{R}, \rho) \to (Y, d)$ is a bijective isometry. By referring to the geometry it is easy to see that the sequence $\{n\}$ is a Cauchy sequence in $(\mathbb{R}, \rho)$, but of course it is not in the usual metric for $\mathbb{R}$. Thus, $(\mathbb{R}, \rho)$ is not complete.

Definition 1.3.15. A function $f : (X, d) \to (Z, \rho)$ between two metric spaces is *uniformly continuous* if for every $\epsilon > 0$ there is a δ such that $\rho(f(x), f(y)) < \epsilon$ when $d(x, y) < \delta$.

The reader may have seen this concept in calculus. The difference between a continuous and a uniformly continuous function, as you might have observed, is the following. When given an ϵ, for uniform continuity we need to find one δ that will work for all pairs of points x, y; for continuity we need to find a δ for only one point x at a time. That is, in the definition of continuity the δ depends on the point x as well as ϵ, while for uniform continuity the δ depends only on ϵ. So every uniformly continuous function is continuous, but there are continuous functions that are not uniformly continuous, as we will see in the next example.

Example 1.3.16. (a) A function $f : (X, d) \to (Z, \rho)$ is a *Lipschitz function* if there is a constant $M > 0$ such that $\rho(f(x), f(y)) \le M d(x, y)$ for all x, y in X. A ready collection of examples occurs by letting I be an interval in $\mathbb{R}$ and letting $f : I \to \mathbb{R}$ be a continuously differentiable function with $|f'(x))| \le M$ for all x in I. Thus, $|f(x) - f(y)| = \left| \int_y^x f'(t)dt \right| \le \int_y^x |f'(t)|dt \le M|x - y|$. Every Lipschitz function is uniformly continuous since for any $\epsilon > 0$ we can take $\delta = \epsilon/M$.

(b) We note that when $A \subseteq X$, the function $x \mapsto \mathrm{dist}\,(x, A)$ is a Lipschitz function by Proposition 1.3.5. Thus the distance function gives rise to a plentiful source of uniformly continuous functions on any metric space.

(c) The function $f : \mathbb{R} \to \mathbb{R}$ defined by $f(x) = x^2$ is not uniformly continuous. In fact, for any $\delta > 0$, consider $x = n$, a natural number, and $y = n + \delta$. So $|f(x) - f(y)| = (n + \delta)^2 - n^2 > 2n\delta$, and this can be made as large as desired no matter how small we make δ.

(d) The function $f : (0, 1] \to \mathbb{R}$ defined by $f(t) = \sin(t^{-1})$ is continuous but not uniformly continuous. That it is continuous follows from calculus and writing f as the composition of the two functions $t \mapsto t^{-1}$ and the sine function. The fact that it is not uniformly continuous can be seen as follows. If $\delta > 0$, then there are points s and t in the interval $(0, \delta)$ with $f(s) = 1$ and $f(t) = -1$, so that $f(s) - f(t) = 2$, even though $|s - t| < \delta$.

The proof of part (b) in the next proposition is a good one to truly concentrate on as it has techniques that will surface many times in your future.

Proposition 1.3.17. (a) *If $f : (X, d) \to (Z, \rho)$ is a uniformly continuous function and $\{x_n\}$ is a Cauchy sequence in X, then $\{f(x_n)\}$ is a Cauchy sequence in Z.*

(b) *If $A \subseteq X$, (Z, ρ) is a complete metric space, and $f : A \to Z$ is uniformly continuous, then f can be extended to a uniformly continuous function $f : \mathrm{cl}\,A \to Z$.*

Proof.   (a) Let $\epsilon > 0$; thus, there is a $\delta > 0$ such that $\rho(f(x), f(y)) <$
ϵ whenever $d(x, y) < \delta$. If $\{x_n\}$ is a Cauchy sequence in X, then
let N be chosen such that $d(x_n, x_m) < \delta$ when $m, n \geq N$. Thus,
$\rho(f(x_n), f(x_m)) < \epsilon$ when $m, n \geq N$, and therefore $\{f(x_n)\}$ is a Cauchy
sequence.

(b) If $x \in \operatorname{cl} A$, then there is a sequence $\{a_n\}$ in A such that $a_n \to x$.
By part (a) this implies that $\{f(a_n)\}$ is a Cauchy sequence in (Z, ρ);
since (Z, ρ) is complete, there is a z in Z such that $f(a_n) \to z$. We
want to define $f(x) = z$; to do this, we need to show that the point
z is independent of which sequence $\{a_n\}$ we choose to converge to x—
otherwise, setting $f(x) = z$ is an ambiguous definition. So assume that
$\{b_n\}$ is also a sequence in A that converges to x and let $\zeta \in Z$ such
that $f(b_n) \to \zeta$; we want to show that $\zeta = z$. Observe that if we define
$c_{2n} = a_n$ and $c_{2n+1} = b_n$, then $c_n \to x$; let $y = \lim_n f(c_n)$. Since $\{f(c_n)\}$
has a subsequence that converges to z and another that converges to ζ,
it must be that $z = y = \zeta$. Thus $f : \operatorname{cl} A \to Z$ is well defined if we set
$f(x) = \lim_n f(a_n)$ whenever $\{a_n\}$ is any sequence in A such that $a_n \to x$.
Now we will show that this extended function is uniformly continuous.

Let $\epsilon > 0$, and choose $\delta > 0$ such that $\rho(f(a), f(b)) < \frac{1}{3}\epsilon$ whenever
$a, b \in A$ and $d(a, b) < \delta$. Assume $x, y \in \operatorname{cl} A$ and $d(x, y) < \frac{1}{3}\delta$. So there
are sequences $\{a_n\}$ and $\{b_n\}$ of points in A such that $a_n \to x$ and $b_n \to y$.
Let $N_1 \geq 1$ such that $\rho(f(a_n), f(x)) < \frac{1}{3}\epsilon$ and $\rho(f(b_n), f(y)) < \frac{1}{3}\epsilon$ when
$n \geq N_1$. Thus for $n \geq N_1$

$$\rho(f(x), f(y)) \leq \rho(f(x), f(a_n)) + \rho(f(a_n), f(b_n)) + \rho(f(b_n), f(y))$$

$$< \frac{2\epsilon}{3} + \rho(f(a_n), f(b_n)).$$

Now we can choose $N_2 \geq 1$ such that when $n \geq N_2$, we have $d(a_n, x) <$
$\frac{1}{3}\delta$ and $d(b_n, y) < \frac{1}{3}\delta$. (Why?) Thus, $d(a_n, b_n) \leq d(a_n, x) + d(x, y) +$
$d(y, b_n) < \delta$. But since $a_n, b_n \in A$, this implies $\rho(f(a_n), f(b_n)) < \frac{1}{3}\epsilon$
whenever $n \geq N_2$. From the preceding inequality with $n \geq N =$
$\max\{N_1, N_2\}$ we have that $\rho(f(x), f(y)) < \epsilon$ whenever $x, y \in \operatorname{cl} A$ and
$d(x, y) < \frac{1}{3}\delta$. Therefore, $f : \operatorname{cl} A \to Z$ is uniformly continuous. ∎

The preceding proposition underlines an important distinction between
continuous and uniformly continuous functions. We saw in Example 1.3.14
that two equivalent metrics do not necessarily have the same Cauchy sequences, so this shows in dramatic fashion that Proposition 1.3.17(a) fails for
continuous functions. If we consider the metric space $X = (0, 1]$ and the continuous function $f : (0, 1] \to \mathbb{R}$ defined by $f(x) = \sin(x^{-1})$, then we see that
part (b) of the preceding proposition fails if we only assume that the function
is continuous (Exercise 1). This also shows that $\sin(x^{-1})$ is not uniformly
continuous, though we gave a more direct proof of this in Example 1.3.16.

We might also mention that the converse of Proposition 1.3.17(a) does not hold as the function $f(x) = x^2$ maps Cauchy sequences to Cauchy sequences, but, as we have seen, it is not uniformly continuous (Exercise 10).

Exercises

(1) Show that for any number a the function $f : [0,1] \to \mathbb{R}$ defined by $f(x) = \sin(x^{-1})$ when $x \neq 0$ and $f(0) = a$ is not continuous at 0.

(2) If (X, d) is a metric space, $f : B(a; r) \to \mathbb{R}$ is continuous at a with $f(a) = 0$, and $g : B(a; r) \to \mathbb{R}$ is a bounded function (but not necessarily continuous), then fg is continuous at a.

(3) If $f : (X, d) \to (Z, \rho)$ is continuous, A is a dense subset of X, and $z \in Z$ such that $f(a) = z$ for every a in A, show that $f(x) = z$ for every x in X.

(4) If $f : (X, d) \to (Z, \rho)$ is both continuous and surjective and A is a dense subset of X, show that $f(A)$ is a dense subset of Z.

(5) In Theorem 1.3.3, give an independent proof that shows that conditions (a) and (c) are equivalent. (Here, "independent" means that the proof should not use the equivalence of (a) and (b) or of (b) and (c).)

(6) Prove the statement made in Example 1.3.4(d). What about a continuous function from $[0,1] \cup [2,3]$ into X?

(7) Let q and p be natural numbers, and show that the metric on $\mathbb{R}^{q+p}$ is equivalent to the metric it has if we identify $\mathbb{R}^{q+p}$ with $\mathbb{R}^q \times \mathbb{R}^p$.

(8) If (X_1, d_1) and (X_2, d_2) are two metric spaces, show that the map $\pi_1 : X_1 \times X_2 \to X_1$ defined by $\pi_1(x_1, x_2) = x_1$ is continuous.

(9) If $(X, d), (Z_1, \rho_1), (Z_2, \rho_2)$ are metric spaces and for $k = 1, 2$, $\pi_k : Z_1 \times Z_2 \to Z_k$ is the projection $\pi_k(z_1, z_2) = z_k$, show that a map $f : X \to Z_1 \times Z_2$ is continuous if and only if $\pi_k \circ f$ is continuous for $k = 1, 2$.

(10) Show that x^2 maps Cauchy sequences in $\mathbb{R}$ into Cauchy sequences. (Hint: use Proposition 1.2.16(c).)

(11) (a) Give an example of two equivalent metrics on a set X that have different sets of uniformly continuous functions. (b) Do the equivalent metrics in Exercise 1.1.11 have the same uniformly continuous functions?

(12) Is the composition of two uniformly continuous functions a uniformly continuous function?

(13) Note that Proposition 1.3.7 says that if (X, d) is a metric space and $C(X)$ is the set of all continuous functions $f : X \to \mathbb{R}$, then $C(X)$ is a vector space over $\mathbb{R}$. Show that $C(X)$ is a finite-dimensional vector space if and only if X is a finite set. (Hint: use Urysohn's Lemma.)

(14) Let (X, d) be a metric space, and let $\mathcal{U}$ denote the set of all uniformly continuous functions from X into $\mathbb{R}$. (a) If $f, g \in \mathcal{U}$ and we define $(f + g) : X \to \mathbb{R}$ by $(f + g)(x) = f(x) + g(x)$ for all x in X, show that $f + g \in \mathcal{U}$. In words, $\mathcal{U}$ is a vector space over $\mathbb{R}$. (b) If $f, g \in \mathcal{U}$ and we define $(fg) : X \to \mathbb{R}$ by $(fg)(x) = f(x)g(x)$ for all x in X,

show by an example that it does not necessarily follow that $fg \in \mathcal{U}$.
If, however, the functions are also bounded, then $fg \in \mathcal{U}$. [A function
$f : X \to (Z, \rho)$ is *bounded* if $f(X)$ is a bounded subset of Z.] (c) Can
you give some conditions under which the quotient of two uniformly
continuous functions is uniformly continuous?

(15) Is the function π_1 defined in Exercise 8 uniformly continuous?

1.4. Compactness

If $\mathcal{G}$ is a collection of subsets of X and $E \subseteq X$, then $\mathcal{G}$ is a *cover* of E if
$E \subseteq \bigcup \{ G : G \in \mathcal{G} \}$. A *subcover* of E is a subset $\mathcal{G}_1$ of $\mathcal{G}$ that is also a cover
of E. Finally, we say that $\mathcal{G}$ is an *open cover* of E if $\mathcal{G}$ is a cover and every
set in the collection $\mathcal{G}$ is open.

Definition 1.4.1. A subset K of the metric space (X, d) is said to be *compact*
if every open cover of K has a finite subcover.

We mention that the term "open cover" in this definition can be replaced
by "cover by subsets of K that are relatively open" (Exercise 2).

It is easy to find examples of sets that are not compact. Specifically,
the open interval $(0, 1)$ is not compact since if we set $G_n = (n^{-1}, 1)$, then
$\mathcal{G} = \{ G_n : n \in \mathbb{N} \}$ is an open cover of the interval that has no finite subcover.
Similarly, $\mathbb{R}$ is not compact since $\{ (-n, n) : n \in \mathbb{N} \}$ is an open cover of
$\mathbb{R}$ that has no finite subcover. We can easily see that every finite subset of
X is compact, but finding nontrivial examples of compact sets requires us to
first prove some results.

Proposition 1.4.2. *Let (X, d) be a metric space.*

(a) *If K is a compact subset of X, then K is closed and bounded.*

(b) *If K is compact and F is a closed set contained in K, then F is compact.*

(c) *The continuous image of a compact subset is a compact subset.*

Proof. (a) If $x \notin K$, then for each z in K let $r_z, s_z > 0$ such that
$B(z; r_z) \cap B(x; s_z) = \emptyset$. Now $\{ B(z; r_z) : z \in K \}$ is an open cover
of K. Since K is compact, there are points $z_1, \ldots, z_n$ in K such that
$K \subseteq \bigcup_{k=1}^{n} B(z_k; r_{z_k})$. Let $s = \min \{ s_{z_k} : 1 \le k \le n \}$. Note that
$B(x; s) \cap K = \emptyset$; in fact, if there is a y in $B(x; s) \cap K$, then there is a k
such that $y \in B(x; s) \cap B(z_k; r_{z_k}) \subseteq B(x; s_{z_k}) \cap B(z_k; r_{z_k})$, which contradicts the choice of the numbers s_{z_k} and r_{z_k}. Therefore, $B(x; s) \subseteq X \backslash K$.
Since x was arbitrary, this says that $X \backslash K$ is open. Also, for any point
x_0 in X, $\{ B(x_0; n) : n \in \mathbb{N} \}$ is an open cover of K; hence there is a
finite subcover. But the sets in this cover are increasing, so there is an
integer n such that $K \subseteq B(x_0; n)$ and K is bounded.

(b) Let $\mathcal{G}$ be an open cover of F, and observe that since F is closed, $\{ X \backslash F \} \cup$
$\mathcal{G}$ is an open cover of K. The existence of a finite subcover of K implies
there is a finite subcollection of $\mathcal{G}$ that covers F.

(c) Let $f : X \to (Z, \rho)$ be a continuous function, and assume K is a compact subset of X; we want to show that $f(K)$ is a compact subset of Z. Let $\mathcal{U}$ be an open cover of $f(K)$ in (Z, ρ). Since f is continuous, it follows that $\mathcal{G} = \{f^{-1}(U) : U \in \mathcal{U}\}$ is an open cover of K. Therefore, there are sets $U_1, \ldots, U_n$ in $\mathcal{U}$ such that $K \subseteq \bigcup_{k=1}^{n} f^{-1}(U_k)$. It follows that $f(K) \subseteq \bigcup_{k=1}^{n} U_k$. ∎

Now for a result from calculus that is an easy consequence of what we have done. This is sometimes called the extreme value property or theorem.

Corollary 1.4.3. *If (X, d) is a compact metric space and $f : X \to \mathbb{R}$ is a continuous function, then there are points a and b in X such that $f(a) \leq f(x) \leq f(b)$ for all x in X.*

Proof. We have from Proposition 1.4.2 that $f(X)$ is a closed and bounded subset of $\mathbb{R}$. Put $\alpha = \inf\{f(x) : x \subset X\}, \beta = \sup\{f(x) : x \in X\}$. Since $f(X)$ is closed, $\alpha, \beta \in f(X)$ by the Completeness Property (Axiom A.3.3) of $\mathbb{R}$; this proves the corollary. ∎

We need two more definitions.

Definition 1.4.4. Say that a subset K of the metric space (X, d) is *totally bounded* if for any radius $r > 0$ there are points $x_1, \ldots, x_n$ in K such that $K \subseteq \bigcup_{k=1}^{n} B(x_k; r)$. A collection $\mathcal{F}$ of subsets of K has the *finite intersection property* (FIP) if whenever $F_1, \ldots, F_n \in \mathcal{F}$, $\bigcap_{k=1}^{n} F_k \neq \emptyset$.

The following theorem is the main result on compactness in metric spaces.

Theorem 1.4.5. *The following statements are equivalent for a closed subset K of a metric space (X, d).*

(a) *K is compact.*

(b) *If $\mathcal{F}$ is a collection of closed subsets of K having the FIP, then it holds that $\bigcap_{F \in \mathcal{F}} F \neq \emptyset$.*

(c) *Every sequence in K has a convergent subsequence.*

(d) *Every infinite subset of K has a limit point.*

(e) *(K, d) is a complete metric space that is totally bounded.*

Proof. (a) *implies* (b). Let $\mathcal{F}$ be a collection of closed subsets of K having the FIP. Suppose $\bigcap\{F : F \in \mathcal{F}\} = \emptyset$. If $\mathcal{G} = \{X \backslash F : F \in \mathcal{F}\}$, then it follows that $\mathcal{G}$ is an open cover of X and, therefore, of K. By (a), there are $F_1, \ldots, F_n$ in $\mathcal{F}$ such that $K \subseteq \bigcup_{j=1}^{n}(X \backslash F_j) = X \backslash \left[\bigcap_{j=1}^{n} F_j\right]$. But since each F_j is a subset of K, this implies $\bigcap_{j=1}^{n} F_j = \emptyset$, contradicting the fact that $\mathcal{F}$ has the FIP.

(b) *implies* (a). Let $\mathcal{G}$ be an open cover of K, and put $\mathcal{F} = \{K \backslash G : G \in \mathcal{G}\}$. Since $\mathcal{G}$ covers K, $\bigcap\{K \backslash G : G \in \mathcal{G}\} = \emptyset$. Thus $\mathcal{F}$ cannot have the FIP and there must be a finite number of sets $G_1, \ldots, G_n$ in $\mathcal{G}$ with $\emptyset = \bigcap_{j=1}^{n}(K \backslash G_j)$. But this implies that $\{G_1, \ldots, G_n\}$ is a finite cover of K. Hence K is compact.

(d) *implies* (c). Assume $\{x_n\}$ is a sequence of distinct points in K. By (d), $\{x_n\}$ has a limit point; since K is closed, that limit point must be in K. We are tempted here to invoke Proposition 1.2.7(a), but we must manufacture an actual subsequence of the original sequence. This takes a little bit of care and effort, which we leave to the interested reader.

(c) *implies* (d). If S is an infinite subset, then S has a sequence of distinct points $\{x_n\}$; by (c), there is a subsequence $\{x_{n_k}\}$ that converges to some point x. It follows that x is a limit point of S. (Details?)

(a) *implies* (d). Assume that (d) is false. So there is an infinite subset S of K with no limit point; it follows that there is an infinite sequence $\{x_n\}$ in S with no limit point. Thus, for each $n \geq 1$, $F_n = \{x_k : k \geq n\}$ contains all its limit points and is therefore closed. Also, $\bigcap_{n=1}^{\infty} F_n = \emptyset$. But each finite subcollection of $\{F_1, F_2, \ldots\}$ has nonempty intersection, contradicting (b), which is equivalent to (a).

(a) *implies* (e). First, let $\{x_n\}$ be a Cauchy sequence in K. Since (a) implies (d), which is equivalent to (c), there is an x in K and a subsequence $\{x_{n_k}\}$ such that $x_{n_k} \to x$. But this implies $x_n \to x$ by Proposition 1.2.11. Hence, (K, d) is complete. To show that K is totally bounded, just note that $\{B(x; r) : x \in K\}$ is an open cover of K for any $r > 0$.

(e) *implies* (c). Fix an infinite sequence $\{x_n\}$ in K, and let $\{\epsilon_n\}$ be a decreasing sequence of positive numbers such that $\epsilon_n \to 0$. By (e), there is a covering of K by a finite number of balls of radius ϵ_1. Thus there is a ball $B(y_1; \epsilon_1)$ that contains an infinite number of points from $\{x_n\}$; let $\mathbb{N}_1 = \{n \in \mathbb{N} : d(x_n, y_1) < \epsilon_1\}$. Now consider the sequence $\{x_n : n \in \mathbb{N}_1\}$ and balls of radius ϵ_2. As we just did, there is a point y_2 in K such that $\mathbb{N}_2 = \{n \in \mathbb{N}_1 : d(y_2, x_n) < \epsilon_2\}$ is an infinite set. Using induction we can show that for each $k \geq 1$ we get a point y_k in K and an infinite set of positive integers $\mathbb{N}_k$ such that $\mathbb{N}_{k+1} \subseteq \mathbb{N}_k$ and $\{x_n : n \in \mathbb{N}_k\} \subseteq B(y_k; \epsilon_k)$. If $F_k = \mathrm{cl}\,\{x_n : n \in \mathbb{N}_k\}$, then $F_{k+1} \subseteq F_k$ and $\mathrm{diam}\, F_k \leq 2\epsilon_k$. Since K is complete, Cantor's Theorem implies that $\bigcap_{k=1}^{\infty} F_k = \{x\}$ for some point x in X. Now, using a small induction argument, pick integers n_k in $\mathbb{N}_k$ such that $n_k < n_{k+1}$. It follows that $\{x_{n_k}\}$ is a subsequence of the original sequence and $x_{n_k} \to x$.

(e) *implies* (a). We first prove the following claim.

Claim 1.4.6. If K satisfies (c) and $\mathcal{G}$ is an open cover of K, then there is an $r > 0$ such that for each x in K there is a G in $\mathcal{G}$ such that $B(x; r) \subseteq G$.

Let $\mathcal{G}$ be an open cover of K, and suppose the claim is false; thus, for every $n \geq 1$ there is an x_n in K such that $B(x_n; n^{-1})$ is not contained in any set G in $\mathcal{G}$. By (c), there is an x in K and a subsequence $\{x_{n_k}\}$ such that $x_{n_k} \to x$. Since $\mathcal{G}$ is a cover, there is a G in $\mathcal{G}$ such that $x \in G$; choose a positive ϵ such that $B(x; \epsilon) \subseteq G$. Let $n_k > 2\epsilon^{-1}$ such that $x_{n_k} \in B(x; \epsilon/2)$. If $y \in B(x_{n_k}; n_k^{-1})$, then $d(x, y) \leq d(x, x_{n_k}) + d(x_{n_k}, y) < \epsilon/2 + n_k^{-1} < \epsilon$, so that $y \in B(x; \epsilon) \subseteq G$. Thus $B(x_{n_k}; n_k^{-1}) \subseteq G$, contradicting the restriction imposed on x_{n_k}. This establishes the claim.

From here it is easy to complete the proof. We know that (e) implies (c), so for an open cover $\mathcal{G}$ of K let $r > 0$ be the number guaranteed by Claim 1.4.6. Now let $x_1, \ldots, x_n \in K$ such that $K \subseteq \bigcup_{k=1}^{n} B(x_k; r)$, and for $1 \le k \le n$ let $G_k \in \mathcal{G}$ such that $B(x_k; r) \subseteq G_k$. $\{G_1, \ldots, G_n\}$ is the sought-after finite subcover.

(c) *implies* (e). If $\{x_n\}$ is a Cauchy sequence in K, then (c) implies it has a convergent subsequence; by Proposition 1.2.11, the original sequence converges. Thus (K, d) is complete. Now fix an $r > 0$. Let $x_1 \in K$; if $K \subseteq B(x_1; r)$, then we are done. If not, then there is a point x_2 in $K \backslash B(x_1; r)$. Once again, if $K \subseteq B(x_1; r) \cup B(x_2; r)$, we are done; otherwise, pick an x_3 in $K \backslash [B(x_1; r) \cup B(x_2; r)]$. Continue. If this process does not stop after a finite number of steps, then we produce an infinite sequence $\{x_n\}$ in K with $d(x_n, x_m) \ge r$ whenever $n \ne m$. But this implies that this sequence can have no convergent subsequence, contradicting (c). ∎

Compactness is one of the most important properties; many good things follow from it, as we shall see in this book, and readers will continue to see as they continue their careers. The point is that compact sets are *almost* finite in a very technical sense, and this approximation of finiteness often suffices to allow us to carry out an argument for a compact set that we can easily make for a finite set. We already saw this in Corollary 1.4.3.

The next result gives the converse of Proposition 1.4.2(a) for $\mathbb{R}^q$, but first we need a lemma.

Lemma 1.4.7. *If* $-\infty < a < b < \infty$ *in* $\mathbb{R}$, *then* $[a, b]$ *is compact.*

Proof. Clearly a closed interval, being a closed subset of a complete metric space, is complete. To show $[a, b]$ is compact, it therefore suffices to show it is totally bounded [Theorem 1.4.5(e)]. But if $r > 0$, then we can easily find $a = x_1 < x_2 < \cdots < x_n = b$ such that $x_{k-1} - x_k < r$, and from here it is routine to show that $[a, b] \subseteq \bigcup_{k=1}^{n} B(x_k; r)$. ∎

Theorem 1.4.8 (Heine[5]–Borel[6] Theorem). *A subset of* $\mathbb{R}^q$ *is compact if and only if it is closed and bounded.*

---

[5]Heinrich Eduard Heine was born in 1821 in Berlin, the eighth of nine children. He received his doctorate in 1842 from Berlin; in 1844 he received the habilitation at Bonn, where he was appointed a privatdozent. In 1856 he was made professor at the University of Halle, where he remained for the rest of his career. In 1850 he married Sophie Wolff from Berlin, and over the years they had five children. He worked on partial differential equations and then on special functions—Legendre polynomials, Lamé functions, and Bessel functions. He made significant contributions to spherical harmonics and introduced the concept of uniform continuity. It was in 1872 that he gave a proof of the present theorem, and it requires scholarship to discover the difference in contribution to this result between him and Borel, who published it in 1895. He died in 1881 in Halle.

[6]Emile Borel was born in Saint Affrique in the south of France in 1871. He published his first two papers in 1890, 2 years before receiving his doctorate in Paris and joining the faculty at Lille. He returned to Paris in 1897. In 1909 a special Chair in the Theory of Functions was created for him at the Sorbonne. During World War I he was very supportive of his country and was put in charge of a central department of research. He also spent time at the front, and in 1918 he was awarded the Croix de Guerre. In 1928 he set up the Institute Henri Poincaré. He was one of the founders of the modern theory of functions, along with Baire and Lebesgue, and he also worked on divergent series, complex variables, probability, and game theory. He continued to be

Proof. If K is compact, then K is closed and bounded by Proposition 1.4.2. Now assume that K is closed and bounded. It follows that there are bounded intervals $[a_1, b_1], \ldots, [a_q, b_q]$ in $\mathbb{R}$ such that $K \subseteq [a_1, b_1] \times \cdots \times [a_q, b_q]$. Suppose $\{x_n\}$ is a sequence in K with $x_n = (x_n^1, \ldots, x_n^q)$. Thus, $\{x_n^1\}$ is a sequence in $[a_1, b_1]$, so the preceding lemma implies it has a convergent subsequence. The notation in this proof could become grotesque if we do the standard things, so we depart from the standard. Denote the convergent subsequence by $\{x_n^1 : n \in \mathbb{N}_1\}$, where $\mathbb{N}_1 \subseteq \mathbb{N}$ and $\mathbb{N}_1$ has its natural ordering. We have that the limit exists, so put $x^1 = \lim_{n \in \mathbb{N}_1} x_n^1$. Now consider the sequence $\{x_n^2 : n \in \mathbb{N}_1\}$ in $[a_2, b_2]$. It has a convergent subsequence $\{x_n^2 : n \in \mathbb{N}_2\}$ with $x^2 = \lim_{n \in \mathbb{N}_2} x_n^2$. Continue, and we get $\mathbb{N}_q \subseteq \cdots \subseteq \mathbb{N}_1 \subseteq \mathbb{N}$ and $x^k = \lim_{n \in \mathbb{N}_k} x_n^k$ for $1 \leq k \leq q$. It follows that $\{x_n : n \in \mathbb{N}_q\}$ is a subsequence of the original sequence and it converges to $x = (x^1, \ldots, x^q)$; it must be that $x \in K$ since K is closed. By Theorem 1.4.5, K is compact. ∎

Example 1.4.9. For the metric space $\mathbb{Q}$, if $a, b \in \mathbb{Q}, a < b$, then the set $F = \{x \in \mathbb{Q} : a \leq x \leq b\} = \mathbb{Q} \cap [a, b]$ is closed and bounded but not compact. In fact, there is a sequence $\{x_n\}$ in F that converges to an irrational number, and so this sequence has no subsequence that converges to a point of F. Thus, in an arbitrary metric space, a closed and bounded set need not be compact, and the Heine–Borel Theorem is particular to Euclidean space.

Note that the metric used for $\mathbb{R}^q$ in the Heine–Borel Theorem must be the standard one, or the result may fail even for an equivalent metric. For example, if we use the metric $d(x, y) = |x - y|(1 + |x - y|)^{-1}$ on $\mathbb{R}$, then all of $\mathbb{R}$ is closed and bounded but not compact. This furnishes another example of a closed and bounded set that is not compact.

Theorem 1.4.10. *If (X, d) is a compact metric space and $f : X \to (Z, \rho)$ is a continuous function, then f is uniformly continuous.*

Proof. Suppose f is not uniformly continuous. So there is an $\epsilon > 0$ such that for every $\delta > 0$ there are points x and y in X with $d(x, y) < \delta$, but $\rho(f(x), f(y)) \geq \epsilon$. In particular, for each natural number n there are points x_n, y_n in X such that $d(x_n, y_n) < n^{-1}$, but $\rho(f(x_n), f(y_n)) \geq \epsilon$. Since X is compact, there is a point x in X and a subsequence $\{x_{n_j}\}$ such that $x_{n_j} \to x$. Since $d(y_{n_j}, x) \leq n_j^{-1} + d(x_{n_j}, x)$, we have that $y_{n_j} \to x$. Therefore, since f is a continuous function,

$$\epsilon \leq \rho(f(x_{n_j}), f(y_{n_j})) \leq \rho(f(x_{n_j}), f(x)) + \rho(f(x), f(y_{n_j})) \to 0.$$

This contradiction proves the theorem. ∎

Proposition 1.4.11. *A compact metric space is separable.*

very active in the French government, serving in the French Chamber of Deputies (1924–1936) and as Minister of the Navy (1925–1940). He died in 1956 in Paris.

Proof. For each natural number n we can find a finite set F_n such that $X = \bigcup \{B(x; n^{-1}) : x \in F_n\}$. Put $F = \bigcup_{n=1}^{\infty} F_n$; we will show that this countable set F is dense in X. In fact, if x_0 is an arbitrary point in X and $\epsilon > 0$, then choose n such that $n^{-1} < \epsilon$. Thus, there is a point x in $F_n \subseteq F$ with $d(x_0, x) < n^{-1} < \epsilon$, proving that $x_0 \in \operatorname{cl} F$. ∎

Exercises

(1) Show that the union of a finite number of compact sets is compact.

(2) If K is a subset of (X, d), show that K is compact if and only if every cover of K by relatively open subsets of K has a finite subcover.

(3) Show that the closure of a totally bounded set is totally bounded.

(4) Show that a totally bounded set is bounded. Is the converse true?

(5) If $\{E_n\}$ is a sequence of totally bounded sets such that $\operatorname{diam} E_n \to 0$, show that $\bigcup_{n=1}^{\infty} E_n$ is totally bounded.

(6) If (X, d) is a complete metric space and $E \subseteq X$, show that E is totally bounded if and only if $\operatorname{cl} E$ is compact.

(7) (a) If G is an open set and K is a compact set with $K \subseteq G$, show that there is a $\delta > 0$ such that $\{x : \operatorname{dist}(x, K) < \delta\} \subseteq G$. (b) Find an example of an open set G in a metric space X and a closed, noncompact subset F of G such that there is no $\delta > 0$ with $\{x : \operatorname{dist}(x, F) < \delta\} \subseteq G$.

(8) If $(X_1, d_1), (X_2, d_2)$ are metric spaces, show that $X_1 \times X_2$ is compact if and only if both X_1 and X_2 are compact.

(9) Give an example of a noncompact metric space (X, d) and a continuous function $f : (X, d) \to (Z, \rho)$ such that $f(X)$ is compact.

(10) For two subsets A and B of X, define the distance from A to B by $\operatorname{dist}(A, B) = \inf\{d(a, b) : a \in A, b \in B\}$. (a) Show that $\operatorname{dist}(A, B) = \operatorname{dist}(B, A) = \operatorname{dist}(\operatorname{cl} A, \operatorname{cl} B)$. (b) If A and B are two disjoint closed subsets of X such that B is compact, then $\operatorname{dist}(A, B) > 0$. (c) Give an example of two disjoint closed subsets A and B of the plane $\mathbb{R}^2$ such that $\operatorname{dist}(A, B) = 0$. (d) Is this exercise related to Exercise 7?

(11) Consider the metric space ℓ^{∞} (see Exercise 1.1.12) and show that

$$\left\{ x = \{x_n\} \in \ell^{\infty} : \sup_n |x_n| \le 1 \right\}$$

is not totally bounded and, therefore, not compact.

(12) Say that a metric space is σ-*compact* if it can be written as the union of a countable number of compact sets. (a) Give three examples of σ-compact metric spaces that are not compact. (b) Show that a σ-compact metric space is separable.

1.5. Connectedness

In this section, we introduce and explore another important concept for metric spaces.

Consider the following two examples of subsets of $\mathbb{R}$. The first is the set $X = [0, 1] \cup (2, 3)$, and the second is $Y = [0, 1] \cup (1, 2)$. In X there are two distinct "parts," $[0, 1]$ and $(2, 3)$. In the second we have written Y as the union of two sets, but these two sets are not really separate "parts." (The term "parts" will be made technically precise soon, though we will not use that term.) In a sense, writing Y as the union of those two sets is just accidental. We could just as well have written $Y = [0, 1) \cup [1, 2)$ or $Y = [0, \frac{1}{2}] \cup (\frac{1}{2}, 2)$ or even $Y = [0, 2)$. What is the true difference between the two sets X and Y?

Note that in the metric space (X, d) of the last paragraph, the set $[0, 1]$ is simultaneously both an open and a closed set. For example, $B_X(1; \frac{1}{2}) = \{x \in X : |x - 1| < \frac{1}{2}\} = (\frac{1}{2}, 1] \subseteq [0, 1]$. It follows that $[0, 1]$ is open in X as well as closed; similarly, $(2, 3)$ is open in X as well as closed. The set X is an example of what we now define as a set that is not connected or, more succinctly, a disconnected set.

Definition 1.5.1. A metric space (X, d) is *connected* if there are no subsets of X that are simultaneously open and closed other than X and $\emptyset$. If $E \subseteq X$, we say that E is connected if (E, d) is connected. If E is not connected, then we will say that it is *disconnected* or a *nonconnected* set.

An equivalent formulation of connectedness is to say that (X, d) is connected provided that when $X = A \cup B$, where $A \cap B = \emptyset$ and both A and B or open (or closed), then either $A = \emptyset$ or $B = \emptyset$. This is the sense of our use of the term "parts" in the introduction of this section; for the set X there, $A = [0, 1]$ and $B = (2, 3)$ are two disjoint, nontrivial sets that are both open and closed in X.

The first result can be considered an example, but it is much more than that.

Proposition 1.5.2. *A subset of $\mathbb{R}$ is connected if and only if it is an interval.*

Proof. Assume that $X = [a, b]$, and let us show that X is connected. (The proof that other types of intervals are connected is Exercise 1.) Assume that $[a, b] = A \cup B$, where both A and B are open and $A \cap B = \emptyset$. One of these sets contains the point a; suppose $a \in A$. Note that A is also closed. We want to show that $A = X$. Since A is open, there is an $\epsilon > 0$ such that $[a, a + \epsilon) \subseteq A$. Put $r = \sup\{\epsilon : [a, a + \epsilon) \subseteq A\}$. We claim that $[a, a + r) \subseteq A$. In fact, if $a \le x < a + r$, then the definition of the supremum implies there is an $\epsilon > 0$ such that $\epsilon < r$, $[a, a + \epsilon) \subseteq A$, and $a \le x < a + \epsilon$; thus $x \in A$. Now A is also closed and $[a, a + r) \subseteq A$, so it must also be that $a + r \in A$. If $a + r \ne b$, then the fact that A is open implies that there is a $\delta > 0$ such that $(a + r - \delta, a + r + \delta) \subseteq A$. But this means that $[a, a + r + \delta) = [a, a + r) \cup (a + r - \delta, a + r + \delta) \subseteq A$, contradicting the definition of r. Therefore, $a + r = b$, and we have that $A = [a, b] = X$. (So $B = \emptyset$.)

Let us begin the converse by deciding how we recognize an interval: a subset E of $\mathbb{R}$ is an interval if and only if when $a, b \in E$ and $a < b$, then $c \in E$ whenever $a < c < b$; equivalently, $[a, b] \subseteq E$ when $a, b \in E$ (Exercise 2). So assume that X is a nonempty connected subset of $\mathbb{R}$ and $a, b \in X$ with $a < b$; suppose $a < c < b$. If $c \notin X$, then let $A = X \cap (-\infty, c), B = X \cap (c, \infty)$.

Clearly A and B are open subsets relative to X; since $c \notin X$, we also have that $A = X \cap (-\infty, c], B = X \cap [c, \infty)$, and they are also closed relative to X (Exercise 1.1.6). Since $a \in A$ and $b \in B$, neither is empty, contradicting the assumption that X is connected. ∎

Theorem 1.5.3. *The continuous image of a connected set is connected.*

Proof. Let $f : (X, d) \to (Z, \rho)$ be a continuous function and E a connected subset of X; we want to show that $f(E)$ is a connected subset of Z. By replacing X with E, we may assume $X = E$ is connected; by replacing Z with $f(E)$, we may assume that f is surjective. We must now show that Z is connected. If D is a subset of Z that is both open and closed, then the continuity of f implies $f^{-1}(D)$ is both open and closed in X. Since X is connected, $f^{-1}(D)$ is either $\emptyset$ or X. But since f is surjective, this implies D is either $\emptyset$ or Z. Thus Z is connected. ∎

The preceding theorem allows us to deduce a standard result from calculus, which here is placed in a more general setting.

Corollary 1.5.4 (Intermediate Value Theorem). *If $f : (X, d) \to \mathbb{R}$ is continuous, X is connected, $a, b \in f(X)$ with $a < b$, then for any number c in the interval $[a, b]$ there is a point x in X with $f(x) = c$.*

Proof. We know that $f(X)$ is a connected subset of $\mathbb{R}$ so that it must be an interval (Proposition 1.5.2). Since $a, b \in f(X)$, it must be that $[a, b] \subseteq f(X)$. ∎

Example 1.5.5. (a) If $x, y \in \mathbb{R}^q$, then the straight line segment $[x, y] \equiv \{ty + (1 - t)x : 0 \leq t \leq 1\}$ is connected. In fact, $t \mapsto ty + (1 - t)x$ is a continuous function from the unit interval into $\mathbb{R}^q$. Since the unit interval is connected, so is its image under this continuous mapping.

(b) In $\mathbb{R}^q$, the balls $B(x; r)$ are connected. In fact, let A be a nonempty subset of $B(x; r)$ that is both relatively open and closed, and fix a point y in A. If $z \in B(x; r)$, then the line segment $[y, z] \subseteq B(x; r)$ and $[y, z]$ is connected by part (a). But it follows that $A \cap [y, z]$ is a nonempty subset of this line segment that is both relatively open and relatively closed. Thus $[y, z] \subseteq A$; in particular, the arbitrary point z from $B(x; r)$ belongs to A, so that $A = B(x; r)$.

(c) If (X, d) is a discrete metric space, then X is not connected if X has more than one point. In fact, each singleton set $\{x\}$ is a nonempty set that is both open and closed.

Proposition 1.5.6. *Let (X, d) be a metric space.*

(a) *If $\{E_i : i \in I\}$ is a collection of connected subsets of X such that $E_i \cap E_j \neq \emptyset$ for all i, j in I, then $E = \bigcup_{i \in I} E_i$ is connected.*

(b) *If $\{E_n : n \geq 1\}$ is a sequence of connected subsets of X such that $E_n \cap E_{n+1} \neq \emptyset$ for each n, then $E = \bigcup_{n=1}^{\infty} E_n$ is connected.*

Proof. (a) Let A be a nonempty subset of E that is relatively open and closed. If $i \in I$, then $A \cap E_i$ is a relatively closed and open subset of E_i; if $A \cap E_i \neq \emptyset$, then the fact that E_i is connected implies $E_i \subseteq A$. Now since A is nonempty, there is at least one i such that $E_i \subseteq A$. But then for every j in I, the hypothesis implies there is a point in E_j that belongs to A; thus $E_j \subseteq A$. Therefore, $A = E$ and E must be connected.

(b) Let A be a nonempty relatively open and closed subset of E. Since $A \neq \emptyset$, there is some integer N with $A \cap E_N \neq \emptyset$. But $A \cap E_N$ is both relatively open and closed in E_N, so $E_N \subseteq A$ by the connectedness of E_N. By hypothesis, $E_{N-1} \cap E_N \neq \emptyset$, so $E_{N-1} \cap A \neq \emptyset$, and it follows that $E_{N-1} \subseteq A$. Continuing, we get that $E_n \subseteq A$ for $1 \leq n \leq N$. Since $E_N \cap E_{N+1} \neq \emptyset$, similar arguments show that $E_{N+1} \subseteq A$. Continuing, we get that $E_n \subseteq A$ for all $n \geq 1$. That is, $E = A$, and so E is connected. ∎

Corollary 1.5.7. *The union of two intersecting connected subsets of a metric space is connected.*

Definition 1.5.8. If (X, d) is a metric space, then a *component* of X is a maximal connected subset of X.

The word *maximal* in the definition means that there is no connected set that properly contains it. Thus, if C is a component of X and D is a connected subset of X with $C \subseteq D$, then $D = C$.

A component is the correct interpretation of the word *part* used in the introduction of this section. The set X there has two components. Notice that a connected metric space has only one component. In a discrete metric space, each singleton set is a component.

Proposition 1.5.9. *For any metric space, every connected set is contained in a component, distinct components are disjoint, and the union of all the components is the entire space.*

Proof. Fix a connected subset D of X, and let $\mathcal{C}_D$ denote the collection of all connected subsets of X that contain D. According to Proposition 1.5.6(a), $C = \bigcup \{A : A \in \mathcal{C}_D\}$ is connected. Clearly, C is a component and contains D. By taking $D = \{x\}$ in what was just established, we have that every point of X is contained in a component so that the union of all the components is X. Finally, note that if C and D are two components and $C \cap D \neq \emptyset$, then $C \cup D$ is connected by Corollary 1.5.7; so it must be that $C = C \cup D = D$ by the maximality of C and D. That is, distinct components are disjoint sets. ∎

One consequence of the preceding proposition is that the components form a partition of X—they divide the space X into a collection of pairwise disjoint connected sets. The next result says that the components are all closed, the proof of which emphasizes once again that, when discussing relatively open and closed sets, you must be aware of what the universe is.

Proposition 1.5.10. *If C is a connected subset of the metric space X and $C \subseteq Y \subseteq \operatorname{cl} C$, then Y is connected.*

Proof. Let A be a nonempty subset of Y that is both relatively open and closed, and fix a point x_0 in A. By Exercise 1.1.6, there is an open subset G of X such that $A = Y \cap G$. Since $x_0 \in \operatorname{cl} C$ and $x_0 \subset G$, there must be a point x in $G \cap C = A \cap C$; that is, $A \cap C$ is a nonempty relatively open subset of C. Since A is relatively closed in Y, Exercise 1.1.6 and an analogous argument imply that $A \cap C$ is also relatively closed in C. Since C is connected, $C = C \cap A \subseteq A$. That is, $C \subseteq A \subseteq Y \subseteq \operatorname{cl} C$, so that A is both closed in Y and dense in Y; hence $A = Y$, and it must be that Y is connected. $\blacksquare$

Corollary 1.5.11. *The closure of a connected set is connected and each component is closed.*

In light of the preceding proposition and Example 1.5.5(b), if $x \in \mathbb{R}^q$ and $B(x;r) \subseteq E \subseteq \overline{B}(x;r)$, then E is connected. Here is an example that will illustrate additional properties as we proceed. In fact this example is used so often it has a name, the *topologist's sine curve*.

Example 1.5.12. $X = \{(x, \sin x^{-1}) \in \mathbb{R}^2 : 0 < x \leq 1\} \cup \{(0,0)\}$ is connected. In fact, $f : (0,1] \to X$ defined by $f(x) = (x, \sin x^{-1})$, is a continuous function, so $C = f((0,1])$ is connected. Since $C \subseteq X \subseteq \operatorname{cl} C$, X is connected by Proposition 1.5.10. The space X consists of the graph of the function $\sin x^{-1}$ for $0 < x \leq 1$ together with the origin. Note that instead of the origin, we could have added to the graph any point or any subset of $\{(0,y) : -1 \leq y \leq 1\}$, and the resulting set would still be connected.

Definition 1.5.13. If E is a subset of X, $x, y \in E$, and $\epsilon > 0$, say that there is an ϵ-*chain* from x to y in E when there is a finite number of points $x_1, \ldots, x_n$ in E such that: (i) for $1 \leq k \leq n$, $B(x_k; \epsilon) \subseteq E$; (ii) for $2 \leq k \leq n$, $x_{k-1} \in B(x_k; \epsilon)$; (iii) $x_1 = x$ and $x_n = y$.

The concept of an ϵ-chain has limited value in an arbitrary metric space. For one thing, note that conditions (i) and (ii) of the definition imply that only points in the interior of E can be linked by an ϵ-chain.

Example 1.5.14. If $r > 0$ and $z \in \mathbb{R}^q$, then for any pair of points x and y in $B(z;r)$ and all sufficiently small ϵ there is an ϵ-chain in $B(z;r)$ from x to y. In fact, observe that the straight line segment from x to y is contained in $B(z;r)$, and using this it is easy to construct the ϵ-chain.

Proposition 1.5.15. *Consider the metric space $\mathbb{R}^q$.*

(a) *If G is an open subset of $\mathbb{R}^q$, then every component of G is open and there are countably many components.*

(b) *An open subset G of $\mathbb{R}^q$ is connected if and only if for any x, y in G there is an $\epsilon > 0$ such that there is an ϵ-chain in G from x to y.*

Proof. (a) If H is a component of G and $x \in H$, choose $r > 0$ such that $B(x; r) \subseteq G$. Since $B(x; r)$ is also connected (Example 1.5.5), Corollary 1.5.7 implies $H \cup B(x; r)$ is connected. Since this is also a subset of G, it follows that $H = H \cup B(x; r)$, so $B(x; r) \subseteq H$, and H is open. Because $\mathbb{R}^q$ is separable, there is a countable dense subset D. Now since each component is open, each component contains an element of D and different components contain different points. If there are uncountably many components, this would show that there is an uncountable subset of D (Why?), which is nonsense.

(b) Assume that G satisfies the stated condition, and let us prove that G is connected. Fix x and let H be the component of G that contains x; we want to show that $H = G$. By part (a), we know that H is open. If $y \in G$, then there is an $\epsilon > 0$ such that there is an ϵ-chain $x_1, \ldots, x_n$ in G from x to y. Since $x_{k-1} \in B(x_{k-1}; \epsilon) \cap B(x_k; \epsilon)$ for $2 \leq k \leq n$, Proposition 1.5.6(b) says $B = \bigcup_{k=1}^n B(x_k; \epsilon)$ is connected. Condition (i) of the definition of an ϵ-chain implies $B \subseteq G$, and so $B \subseteq H$. In particular, $y \in H$. Since y was arbitrary, $H = G$ and G is connected.

Now assume that G is connected. Fix a point x in G, and let

$$D = \{y \in G : \text{ there is an } \epsilon > 0 \text{ and an } \epsilon\text{-chain in } G \text{ from } x \text{ to } y\}$$

The strategy of the proof will be to show that D is both relatively open and closed in G; since it is not empty ($x \in D$), it will then follow that $D = G$, and so G will have been shown to satisfy the condition. If $y \in D$, then let $\epsilon > 0$ and let $x_1, \ldots, x_n$ be an ϵ-chain from x to y. It follows from the definition of an ϵ-chain that $B(y; \epsilon) \subseteq D$. Thus, D is open. Now suppose $z \in G \cap \operatorname{cl} D$—this is the relative closure of D in G. (Why?) Choose $r > 0$ such that $B(z; r) \subseteq G$, so $B(z; r) \cup D \subseteq G$. Since $z \in \operatorname{cl} D$, there is a point y in $B(z; r) \cap D$. Let $x_0 = x, x_1, \ldots, x_n$ be an ϵ-chain from x to y. By Exercise 8, there is an ϵ'-chain from x to y in G whenever $0 < \epsilon' < \epsilon$. Applying this with $0 < \epsilon' < \min\{\epsilon, r\}$, we may assume $\epsilon < r$. Using Example 1.5.14 we see that this implies there is an ϵ-chain in G from x to z. Thus $z \in D$, and so D is relatively closed in G. ∎

We note that (i) of the definition of an ϵ-chain was used to establish that the condition in part (b) was sufficient for connectedness. Without this, the result is false, as we see in the following example.

Example 1.5.16. Let $X = \{(x, y) \in \mathbb{R}^2 : y > x^{-1}\} \cup \{(x, y) \in \mathbb{R}^2 : y < 0\}$. Clearly, X is not connected; in fact, it has two components. It is also easy to see that if $\bar{x} = (x, y)$ with $y < 0$ and $\bar{y} = (w, z)$ with $z > w^{-1}$, then for all sufficiently small ϵ there are points $\bar{x}_1, \ldots, \bar{x}_n$ in X such that for $2 \leq k \leq n$, $\bar{x}_{k-1} \in B(\bar{x}_k; \epsilon)$, and $\bar{x}_1 = \bar{x}$ and $\bar{x}_n = \bar{y}$.

Here is another example.

Example 1.5.17. If $X = \mathbb{Q}$, the set of rational numbers, then $\mathbb{Q}$ is not connected. For example, $A = \mathbb{Q} \cap (\pi, \infty)$ is a subset of $\mathbb{Q}$ that is both open and closed. On the other hand, $\mathbb{Q}$ satisfies the condition stated in Proposition 1.5.15(b). Thus, the assumption there that G is open is essential.

We close this section with a circumspective remark. We did not use sequences in this section. Connectedness is one of the only properties of a metric space I know whose examination never uses the concept of a convergent sequence.

Exercises

(1) Prove that open and half-open intervals are connected, completing the proof of half of Proposition 1.5.2.

(2) Is a proof required of the fact, used in the proof of Proposition 1.5.2, that a subset E of $\mathbb{R}$ is an interval if and only if when $a, b \in E$ and $a < b$, then $c \in E$ whenever $a < c < b$?

(3) If (X, d) is connected and $f : X \to \mathbb{R}$ is a continuous function such that $|f(x)| = 1$ for all x in X, show that f must be constant.

(4) If A is a subset of X, define the *characteristic function* of A as the function $\chi_A : X \to \mathbb{R}$ such that $\chi_A(x) = 1$ when $x \in A$ and $\chi_A(x) = 0$ when $x \notin A$. Show that A is simultaneously open and closed if and only if χ_A is continuous.

(5) Look at the preceding exercise for the definition of the characteristic function on a set X. (a) If A and B are subsets of X, which function is $\chi_A \chi_B$? (b) Which function is $\chi_A + \chi_B$? (c) What is the characteristic function of the empty set?

(6) Let I be any nonempty set and for each i in I let X_i be a copy of $\mathbb{R}$ with the metric $d_i(x, y) = |x - y|$. Let X be the disjoint union of the sets X_i. (That is a verbal description that can be used in any circumstance, but if you want precision, you can say $X = \mathbb{R} \times I$, the cartesian product where I has the discrete metric.) Define a metric d on X by letting d agree with d_i on X_i; and when $x \in X_i$, $y \in X_j$, and $i \neq j$, then $d(x, y) = 1$. (Is this the product metric on $\mathbb{R} \times I$?) (a) Show that d is indeed a metric on X. (b) Show that $\{X_i : i \in I\}$ is the collection of components of X and each of these components is an open subset of X. (c) Show that (X, d) is separable if and only if I is a countable set.

(7) Can you think of any way to generalize Proposition 1.5.6(a) and obtain a theorem whose conclusion is that $E = \bigcup \{E_i : i \in I\}$ is connected.

(8) If E is a subset of $\mathbb{R}^q$, $x, y \in E$, and $\epsilon > 0$ such that there is an ϵ-chain from x to y, show that for any ϵ' with $0 < \epsilon' < \epsilon$ there is an ϵ'-chain from x to y.

(9) A *polygon* $[x, x_1, \ldots, x_{n-1}, y]$ in Euclidean space is the union of straight line segments

$$[x, x_1], [x_1, x_2], \ldots, [x_{n-1}, y].$$

(a) Show that a polygon is a connected subset of $\mathbb{R}^q$. (b) Show that an open subset G of $\mathbb{R}^q$ is connected if and only if for any two points x and y in G there is a polygon $[x, x_1, \ldots, x_{n-1}, y]$ contained in G.

1.6. The Baire Category Theorem

The Baire Category Theorem is a useful result that can prove the existence, in fact profusion, of certain objects. That is an abstract statement that has little meaning unless you see some specific instances. One such instance is that it can be used to show that there is a continuous nowhere differentiable function on an interval in $\mathbb{R}$. In fact, it actually shows that the set of such functions is dense in the metric space $C[0, 1]$. See [5, Theorem 17.8].

It is also used to go from statements that involve one point at a time in a metric space to statements that give a uniform condition. Again, that is rather bare of meaning without a specific instance, but supplying one is beyond the scope of this book. Trust me. If you stick with mathematics a little longer, you will see such things as the Principle of Uniform Boundedness in functional analysis whose proof is a standard application of the Baire Category Theorem. (For example, see [2].) We will encounter a use of the Baire Category Theorem below in Exercise 2.2.6, however, when we explore abstract topological spaces in the next chapter.

Theorem 1.6.1 (Baire[7] Category Theorem). *If (X, d) is a complete metric space and $\{U_n\}$ is a sequence of open subsets of X each of which is dense, then $\bigcap_{n=1}^{\infty} U_n$ is dense.*

Proof. To show that $\bigcap_{n=1}^{\infty} U_n$ is dense, it suffices to show that if G is a nonempty open subset of X, then $G \cap \bigcap_{n=1}^{\infty} U_n \neq \emptyset$. (Why?) Since U_1 is dense and open, there is an open ball $B(x_1; r_1)$ with $r_1 < 1$ such that $\mathrm{cl}\, B(x_1; r_1) \subseteq G \cap U_1$. This is the first step in an induction argument that establishes that for each $n \geq 2$ there is an open ball $B(x_n; r_n)$ with $r_n < n^{-1}$ such that

$$\mathrm{cl}\, B(x_n; r_n) \subseteq B(x_{n-1}; r_{n-1}) \cap U_n \subseteq G.$$

---

[7]René-Louis Baire was born in Paris in 1874. His father was a tailor and his family was poor. He won a scholarship, however, that enabled him to receive an excellent education, and he distinguished himself from the start. Soon he gained admission to the prestigious École Normale Supérieure. After this education and further study in mathematics, he obtained a position as a professor at a lycée and began the research that led him to the introduction of his classification of functions of a real variable. He was awarded another scholarship that enabled him to study in Italy, where he met Volterra. His dissertation on discontinuous functions earned him a doctorate in 1899, and in 1901 he secured a position on the faculty at the University of Montpelier. Throughout his life he suffered from poor health, though this did not keep him from his research. In 1907 he was promoted to a professorship at Dijon, but his health interfered with his teaching duties. In 1914 he requested a leave and traveled to Lausanne, Switzerland. While he was there, World War I began, and he was unable to return to France. His health deteriorated further, depression ensued, and he spent the rest of his life on the shores of Lac Léman in Switzerland. It was there that he received the Chevalier de la Legion d'Honneur, and in 1922 he was elected to the Académie des Sciences. He published significant works on number theory and functions. He died in 1932 at Chambéry near Geneva.

The details of this induction argument are left as Exercise 1. Observe that this implies that when $n > N$,

1.6.2 $\operatorname{cl} B(x_n; r_n) \subseteq B(x_N; r_N) \cap U_N \subseteq G \cap U_N.$

Let N be an arbitrary positive integer. It follows from (1.6.2) that for $n, m \geq N$, $d(x_n, x_m) < 2N^{-1}$, and so $\{x_n\}$ is a Cauchy sequence. Since X is complete, there is an x in X such that $x_n \to x$. The same observation (1.6.2) reveals that when $n > m > N$, $x \in \operatorname{cl} B(x_n; r_n) \subseteq B(x_m; r_m) \subseteq G \cap U_N$. Since N was arbitrary, $x \in G \cap \bigcap_{N=1}^{\infty} U_N$. Thus, this last set is nonempty, and so we have established density. ∎

The next result is the form of the theorem that is used very often and is also referred to as the Baire Category Theorem.

Corollary 1.6.3. *If (X, d) is a complete metric space and $\{F_n\}$ is a sequence of closed subsets such that $X = \bigcup_{n=1}^{\infty} F_n$, then there is an n such that we have* int $F_n \neq \emptyset$.

Proof. Suppose int $F_n = \emptyset$ for each $n \geq 1$, and put $U_n = X \backslash F_n$. It follows that each U_n is open and dense, so the theorem implies that $\bigcap_{n=1}^{\infty} U_n$ is dense. But the hypothesis of this corollary implies that this intersection is empty, a dramatic contradiction. ∎

Why is the word *category* used in the name of this theorem? Please excuse the author while he takes a little time to rant. Mathematics loves tradition; largely this is good and has my support. However, occasionally it adopts a word, makes it a definition, and promulgates it far beyond its usefulness. Such is the case here. There is a concept in topology called a set of the *first category*. Sets that are not of the first category are said to be of the *second category*. What are the definitions? I won't tell you. If you are curious, you can look them up, but it will not be helpful. I will say, however, that using this terminology the Baire Category Theorem says that a complete metric space is of the second category. My objection stems from the fact that this "category" terminology does not convey any sense of what the concept is. There is another pair of terms that is used: *meager* or *thin* and *comeager* or *thick*. These at least convey some sense of what the terms mean. But I see no reason to learn additional terminology. The theorem as stated previously says what it says, end of story—and end of rant.

Exercises

(1) Supply the details of the induction argument used in the proof of Theorem 1.6.1.

(2) Say that a set E in a metric space is *nowhere dense* if int $[\operatorname{cl} E] = \emptyset$. (Why is the term *nowhere* used?) If (X, d) is a complete metric space and $A = \bigcup_{n=1}^{\infty} E_n$, where each E_n is nowhere dense, show that $X \backslash A$ is dense in X.

(3) If (X, d) is a complete metric space and $A_1, A_2, \ldots$ are subsets of X such that int $[\bigcup_{n=1}^{\infty} A_n] \neq \emptyset$, then there is an integer n such that int $A_n \neq \emptyset$.

(4) Show that a closed interval in $\mathbb{R}$ cannot be written as a countable union of closed subsets that are pairwise disjoint.

Chapter 2

Topological Spaces

In this chapter we will abstract certain properties of a metric space and thus create a new structure that occurs frequently in mathematics. Then we will extend to this new structure the notions of a continuous function, compactness, and connectedness. As we progress through this chapter we will frequently use metric spaces to illustrate the new concepts, and readers are encouraged to pursue such issues on their own.

Some good references for this material, as well as for what is in the next chapter, are [4] and [6]. A source of examples and counterexamples besides what is in these references is [11].

2.1. Definitions and Examples

Definition 2.1.1. A *topological space* is a pair of objects $(X, \mathcal{T})$, where X is a set and $\mathcal{T}$ is a collection of subsets of X satisfying the following conditions:

(a) $\emptyset, X \in \mathcal{T}$;

(b) if $\{G_i : i \in I\} \subseteq \mathcal{T}$, then $\bigcup_{i \in I} G_i \in \mathcal{T}$;

(c) if $G_1, \ldots, G_n \in \mathcal{T}$, then $\bigcap_{k=1}^{n} G_k \in \mathcal{T}$.

The collection $\mathcal{T}$ is called the *topology* on X and sets in $\mathcal{T}$ are called *open sets*.

Example 2.1.2. (a) If (X, d) is a metric space and $\mathcal{T}$ denotes the open sets defined by d as in §1.1, then $(X, \mathcal{T})$ is a topological space.

(b) If X is any set and $\mathcal{T} = 2^X$, the collection of all subsets of X, then $(X, \mathcal{T})$ is a topological space. This topology is called the *discrete topology* on X. In fact, this example is a special case of the preceding one if we let d be the discrete metric on X.

(c) If X is any set and $\mathcal{T} = \{\emptyset, X\}$, then $\mathcal{T}$ is a topology on X called the *trivial topology*. This topology does not arise from a metric if X has at least two points. (Why?)

J.B. Conway, *A Course in Point Set Topology*, Undergraduate Texts in Mathematics, DOI 10.1007/978-3-319-02368-7_2,

(d) Let $\{(X_i, \mathcal{T}_i) : i \in I\}$ be a collection of topological spaces, where the
    sets $\{X_i : i \in I\}$ are pairwise disjoint subsets of some larger set that we
    can take to be $X = \bigcup_{i \in I} X_i$. (This idea may strike the reader as a bit
    strange, but we will see more specific examples of this phenomenon as
    we progress.) Let $\mathcal{T} = \{G \subseteq X : G \cap X_i \in \mathcal{T}_i$ for each $i \in I\}$. $(X, \mathcal{T})$ is
    a topological space.

(e) Let X be any set, and let $\mathcal{T}$ be the collection of all subsets G such that
    $X \backslash G$ is finite. It follows that $\mathcal{T}$ is a topology on X called the *cofinal*
    topology. We won't see much of this topology.

(f) If $(X, \mathcal{T})$ is a topological space, $Y \subseteq X$, and $\mathcal{T}_Y \equiv \{Y \cap G : G \in \mathcal{T}\}$,
    then $(Y, \mathcal{T}_Y)$ is a topological space. $\mathcal{T}_Y$ is called the *subspace topology*
    or *relative topology* defined by $\mathcal{T}$ on Y. We note that this is consistent
    with what we did when discussing subspaces of a metric space. That is,
    if (X, d) is a metric space, $\mathcal{T}$ denotes the open sets in X, and $Y \subseteq X$,
    then $\mathcal{T}_Y$ is precisely the set of open subsets of Y obtained by restricting
    the metric d to Y. See Proposition 1.1.8.

Definition 2.1.3. $(X, \mathcal{T})$ is called a *Hausdorff*[1] *space* provided for any pair
of distinct points x, y in X where there are disjoint open sets U, V such that
$x \in U$ and $y \in V$.

It is easy to see that the trivial topological space is not a Hausdorff
space, nor is the cofinal topology on any infinite set, but every metric space
is Hausdorff. In fact, if (X, d) is a metric space, $x \neq y$, and $0 < r < \frac{1}{2} d(x, y)$,
then $B(x; r) \cap B(y; r) = \emptyset$. Topological spaces that are not Hausdorff arise
naturally in certain parts of mathematics. Most mathematicians, however,
will not encounter them very often; analysts never will. When a space fails to
be Hausdorff, pathologies prevail. So we are going to avoid them and settle
on the following agreement.

----

[1]Felix Hausdorff was born in 1868 in Breslau, Germany, which became Wrocław, Poland
after World War II. When Hausdorff was young, his family moved to Leipzig, where he grew
up and was educated. His original interests were in literature and music, but bowing to family
pressure he studied astronomy and obtained a doctorate in 1891. He published four papers in the
subject and obtained a habilitation in 1895. Nevertheless, he continued to pursue his interests in
literature and music. He published his first literary work in 1897 under the name Paul Mongré.
Books followed in philosophy in 1898 and poetry in 1900. Clearly he also worked in mathematics
because in 1902 he was appointed to an extraordinary professorship of mathematics at Leipzig
and turned down the offer of a similar appointment at Göttingen. He continued with literature
and published a farce in 1904 that was produced and was an apparent success. After 1904,
however, his efforts shifted to topology, introducing the concept of a partially ordered set (§ A.4).
In 1910 he went to Bonn, and in 1913 he accepted an ordinary professorship at Greifswalf.
(Hausdorff came from a rich family and had no financial worries, hence the willingness to accept
a lower position.) In 1914 he published his book *Grundzüge der Mengenlehre*, in which he set
out the theory of topological spaces, building on the work of Fréchet. In a sense, this is the start
of point set topology, and the reader can find here the introduction of what we are calling the
Hausdorff property. The book was reprinted several times and is a good place to practice your
mathematical German. He continued to be active until 1935 when, as a Jew, he was forced by
the Nazis to retire. He continued to do research but could not find an outlet for his work in
Germany. He tried to emigrate in 1939 but was unsuccessful. In 1941 he was scheduled to be
sent to a concentration camp, but he managed to avoid this. Bonn University requested that he
and his wife be allowed to remain in their home, and this was granted. He committed suicide
together with his wife and her sister in 1942.

Agreement. *All topological spaces encountered in this book will be assumed to be Hausdorff.*

Incidentally, another term for Hausdorff is a T_2-space. The reason for this term is part of what are called separation axioms or properties. Yes, there is a T_1-space and even a T_0-space as well as T_3 and T_4. It is a long story. We are going to avoid this terminology; it just strikes me as not conveying anything except, possibly, a hierarchical code. Nevertheless, the underlying concepts will be seen subsequently in Sects. 3.2 and 3.3. If you are interested in a more thorough inspection of these properties, an examination of [4] or [6] will satisfy.

The reader may have expected this definition: a subset F of $(X, \mathcal{T})$ is *closed* if $X \backslash F \in \mathcal{T}$. Again, we have taken the definition straight from metric spaces.

Proposition 2.1.4. *Let $(X, \mathcal{T})$ be a topological space, and let $\mathcal{F}$ denote the collection of closed subsets of X.*

(a) $\emptyset, X \in \mathcal{F}$.

(b) *If $\{F_i : i \in I\} \subseteq \mathcal{F}$, then $\bigcap_{i \in I} F_i \in \mathcal{F}$.*

(c) *If $F_1, \ldots, F_n \in \mathcal{F}$, then $\bigcup_{k=1}^{n} F_k \in \mathcal{F}$.*

(d) *$\{x\} \in \mathcal{F}$ for every x in X.*

The proof of this is rather straightforward and left to the reader, though we will mention that the proof of (d) uses the Hausdorff property, whereas the proofs of the first three do not.

We continue to use the metric space development from the preceding chapter as a guide for developing the theory of topological spaces. This will be our pattern for the near future. When a proof of something for a topological space is similar to the proof of the corresponding result in metric spaces, we will not present it but refer to its metric space counterpart. Here we begin with a verbatim definition.

Definition 2.1.5. Let $(X, \mathcal{T})$ be a topological space, and let A be a subset of X. The *interior* of A, denoted by int A, is the set defined by int $A = \bigcup \{G : G$ is open and $G \subseteq A\}$. The *closure* of A, denoted by cl A, is the set defined by cl $A = \bigcap \{F : F$ is a closed set and $A \subseteq F\}$. The *boundary* of A, denoted by ∂A, is the set defined by $\partial A = \text{cl } A \cap \text{cl } (X \backslash A)$.

Proposition 2.1.6. *Let $(X, \mathcal{T})$ be a topological space, and assume that $A \subseteq X$.*

(a) *$x \in \text{int } A$ if and only if there is an open set G with $x \in G \subseteq A$.*

(b) *$x \in \text{cl } A$ if and only if for every open set G that contains x we have that $G \cap A \neq \emptyset$.*

(c) *int A is the largest open set contained in A.*

(d) *cl A is the smallest closed set that contains A.*

For the proof of (a) and (b) look at the proof of Proposition 1.1.13. The proof of (c) and (d) is Exercise 5.

Proposition 2.1.7. *Let $(X, \mathcal{T})$ be a topological space, and let A be a subset of X.*

(a) *A is closed if and only if $A = \operatorname{cl} A$.*

(b) *A is open if and only if $A = \operatorname{int} A$.*

(c) *If $A_1, \ldots, A_n$ are subsets of X, then $\operatorname{cl} \left[\bigcup_{k=1}^{n} A_k \right] = \bigcup_{k=1}^{n} \operatorname{cl} A_k$.*

Consult Proposition 1.1.15 and Exercise 1.1.7.

Definition 2.1.8. A subset E of a topological space $(X, \mathcal{T})$ is *dense* if $\operatorname{cl} E = X$. A topological space is *separable* if it has a countable dense subset.

Using Proposition 2.1.6 we easily get the following extension of Proposition 1.1.20.

Proposition 2.1.9. *A set A is dense in $(X, \mathcal{T})$ if and only if for every x in X and every open set G that contains x we have that $G \cap E \neq \emptyset$.*

We also have the corresponding definition of a limit point.

Definition 2.1.10. If $A \subseteq X$, a point x in X is called a *limit point* of A if for every open set G that contains x there is a point a in $G \cap A$ different from x. In other words, $[G \backslash \{x\}] \cap A \neq \emptyset$.

We can define the concept of a convergent sequence in a topological space as follows: $x_n \to x$ in $(X, \mathcal{T})$ if and only if for each open set G containing x there is an N such that $x_n \in G$ for all $n \geq N$. This, however, has only marginal value. For example, the sequential characterization of limit points given in Proposition 1.2.7 is not true in this more general setting. Later we will generalize the idea of a sequence, and this will turn out to have the utility in a topological space that sequences have in a metric space. Right now we want to extend Proposition 1.2.7 to a topological space, but we need to supply a proof because the one given for this result in the metric space setting relies on sequences. A convenient term in a topological space is a *neighborhood* of a point x, which is an open set G that contains x. (By a neighborhood of x some authors mean a set E that contains x in its interior. This has value and convenience, but we will stick with the definition given before this comment.)

Proposition 2.1.11. *Let A be a subset of X.*

(a) *The set A is closed if and only if it contains all its limit points.*

(b) *$\operatorname{cl} A = A \cup \{x : x \text{ is a limit point of } A\}$.*

Proof. Part (a) will follow from (b) if we first show that a limit point of the set of limit points of A is a limit point of A (Exercise 8). So we only prove (b). Let $B = A \cup \{x : x \text{ is a limit point of } A\}$. Assume x is a limit point of

A. If G is a neighborhood of x, then $G \cap A \neq \emptyset$, so by definition $x \in \operatorname{cl} A$. Thus $B \subseteq \operatorname{cl} A$. On the other hand, if $x \in \operatorname{cl} A$, then for every neighborhood G of x, $G \cap A \neq \emptyset$. If x is not a limit point, then for some neighborhood G we have that $G \cap A = \{x\}$, implying $x \in A$. Thus, $\operatorname{cl} A \subseteq B$. ∎

Proposition 2.1.12. *Let $(X, \mathcal{T})$ be a topological space, let Y be a subset of X, and give Y its subspace topology $\mathcal{T}_Y$.*

(a) *A subset A of Y is closed in Y if and only if there is a closed subset F of X such that $A = F \cap Y$.*

(b) *If $A \subseteq Y$, $\operatorname{cl}_Y A$ denotes the closure of A in Y, and $\operatorname{cl}_X A$ denotes the closure of A in X, then $\operatorname{cl}_Y A = Y \cap \operatorname{cl}_X A$.*

(c) *If $A \subseteq Y$, $\operatorname{int}_Y A$ denotes the interior of A in Y, and $\operatorname{int}_X A$ denotes the interior of A in X, then $Y \cap \operatorname{int}_X A \subseteq \operatorname{int}_Y A$.*

Proof. Exercise 9, which also has a comment about part (c). ∎

We close this section with a brief definition to be discussed more fully later.

Definition 2.1.13. A topological space $(X, \mathcal{T})$ is *metrizable* if there is a metric defined on X such that the open sets defined by this metric are precisely the sets $\mathcal{T}$.

Needless to say, every metric space is metrizable, but inherent conditions that are necessary and sufficient for a topological space to be metrizable are a challenge to discover. This will be discussed at the end of § 3.7. There are Hausdorff topological spaces that are not metrizable, but we will have to wait for that as well. In Exercise 2.2.6, an example of a space is given that has both a metric ρ and a topology $\mathcal{T}$, and, with considerable effort, it is shown that the topology $\mathcal{T}$ is not defined by the metric ρ. In fact, that topology $\mathcal{T}$ is not metrizable, but, as far as I know, showing this requires techniques beyond the scope of this book. In fact, mathematics is filled with topological spaces that are not metrizable, but showing this lack of metrizability is involved.

Exercises

(1) Prove that for any topological space X and any point x in X, $\{x\}$ is a closed set.

(2) Verify the statement in Example 2.1.2(d).

(3) Verify the statements made in Example 2.1.2(f).

(4) Prove Proposition 2.1.4(d).

(5) Prove Proposition 2.1.6.

(6) Prove Proposition 2.1.7.

(7) For the moment disregard the agreement at the start of this section that all topological spaces are Hausdorff. Show that a topological space $(X, \mathcal{T})$ is Hausdorff if and only if for any two distinct points x and y there is an open set G such that $y \in G$ and $x \notin \operatorname{cl} G$.

(8) If A is a subset of X and L is the set of limit points of A, prove that any limit point of L is a limit point of A. Is L a closed set?

(9) (a) Prove Proposition 2.1.12. (b) Find an example of a topological space $(X, \mathcal{T})$, a subset Y of X, and a nonempty set U in $\mathcal{T}_Y$ such that int $_X U = \emptyset$. [Thus, we have a dramatic example showing that equality in part (c) of Proposition 2.1.12 does not always occur.]

(10) Let X denote the set of all sequences of real numbers $\{x_n : n \in \mathbb{N}\}$, and let $\mathcal{T}$ consist of all subsets G of X satisfying the following condition: for each $x = \{x_n\}$ in G there are integers $n_1 < \cdots < n_N$ and an $\epsilon > 0$ such that $\{y = \{y_n\} \in X : |x_{n_k} - y_{n_k}| < \epsilon$ for $1 \le k \le N\} \subseteq G$. (a) Show that $(X, \mathcal{T})$ is a topological space. (b) Is there a metric on X such that $\mathcal{T}$ is the collection of open sets for this metric?

2.2. Base and Subbase for a Topology

We want to invent ways to generate a topology on a set. This is the purpose of the present section, and here is the first of two ways we will do this.

Definition 2.2.1. If X is a set, a collection $\mathcal{B}$ of subsets of X is a *base* for a topology $\mathcal{T}$ if every set G in $\mathcal{T}$ is the union of some collection of sets belonging to $\mathcal{B}$.

This is a start, but it is not quite what we are aiming for. We will see the virtues of this concept and a few of its drawbacks and then encounter another way to generate topologies.

Example 2.2.2. (a) If $(X, \mathcal{T})$ is a topological space, then $\mathcal{T}$ is a base for itself.

(b) If (X, d) is a metric space, then $\mathcal{B} = \{B(x; r) : x \in X, r > 0\}$ is a base for the topology on X. So we certainly have that the concept of a base covers our important collection of examples.

(c) If (X, d) is a metric space, then $\mathcal{B} = \{B(x; r) : x \in X, r > 0$ and $r \in \mathbb{Q}\}$ is a base for the topology on X. Thus we see that a base is not unique. Rather than use the positive numbers in $\mathbb{Q}$ we could have only chosen the radii n^{-1} with $n \in \mathbb{N}$; or we could have restricted the possibilities for x so that they come from some prechosen dense set. Thus we have even more varieties of a base. We revisit this in Corollary 2.2.4 below.

The next result presents a base in a form we can use to generate a topology.

Proposition 2.2.3. *If $\mathcal{B}$ is a base for a topology on X, then $\mathcal{B}$ satisfies the following:*

(a) $\bigcup\{B : B \in \mathcal{B}\} = X$;

(b) *if $B_1, B_2 \in \mathcal{B}$ and $x \in B_1 \cap B_2$, then there is a B in $\mathcal{B}$ such that $x \in B \subseteq B_1 \cap B_2$;*

(c) *if x and y are distinct points in X, then there are sets A, B in $\mathcal{B}$ such that $x \in A$, $y \in B$, and $A \cap B = \emptyset$.*

Conversely, if $\mathcal{B}$ is a collection of subsets of X satisfying these three conditions, then $\mathcal{B}$ is a base for a unique Hausdorff topology on X.

Proof. Assume $\mathcal{B}$ is a base for the topology $\mathcal{T}$. Clearly (a) holds since $X \in \mathcal{T}$. Also $\mathcal{B} \subseteq \mathcal{T}$, so when $B_1, B_2 \in \mathcal{B}$, $B_1 \cap B_2 \in \mathcal{T}$ (though possibly not in $\mathcal{B}$) and is thus the union of sets in $\mathcal{B}$; hence (b) holds. If x and y are distinct points in X, then the fact that $(X, \mathcal{T})$ is Hausdorff implies there are neighborhoods G and H of x and y, respectively, such that $G \cap H = \emptyset$. From the definition of a base there are A, B in $\mathcal{B}$ such that $x \in A \subseteq G, y \in B \subseteq H$. Thus (c) holds.

Now for the converse. Assume $\mathcal{B}$ is a collection of subsets of X satisfying the three properties, and let $\mathcal{T}$ be the collection of all subsets of X that are the union of some subcollection of sets from $\mathcal{B}$. Clearly, $X \in \mathcal{T}$, and by taking the union of the sets belonging to the empty subcollection of $\mathcal{B}$ we have that $\emptyset \in \mathcal{T}$. (Do you understand this last statement?) By using condition (b) and doing an induction argument we see that $\mathcal{T}$ is closed under finite intersections. If $\{G_i : i \in I\} \subseteq \mathcal{T}$, then each G_i is the union of sets in $\mathcal{B}$ so that the same holds for $\bigcup_{i \in I} G_i$. Finally, (c) implies that the Hausdorff property holds. Therefore, $\mathcal{T}$ is a topology. An examination of how $\mathcal{T}$ is defined shows that it is the only topology for which $\mathcal{B}$ is a base. (Or we could use Exercise 1.) ∎

When $\mathcal{B}$ satisfies the three conditions of the preceding proposition we call $\mathcal{T}$ the *topology generated by* $\mathcal{B}$. Recall the definition of a separable metric space (Definition 1.1.18).

Corollary 2.2.4. *If (X, d) is separable, $\{a_1, a_2, \dots\}$ is a dense subset of X, and $\{r_1, r_2, \dots\}$ is an enumeration of the rational numbers in the open unit interval, then $\mathcal{B} = \{B(a_n; r_m) : n, m \geq 1\}$ is a base for the topology on X.*

Proof. We show that the three conditions of the preceding proposition are satisfied. Clearly (a) holds. If $x \in B(a_j; r_p) \cap B(a_k; r_q)$, choose an r_m with $0 < r_m < \frac{1}{2} \mathrm{dist}\, [x, X \backslash (B(a_j; r_p) \cap B(a_k; r_q))$. Because $\{a_1, a_2, \dots\}$ is dense, there is an a_n in $B(x; r_m)$. It follows that $B(a_n; r_m) \subseteq B(a_j; r_p) \cap B(a_k; r_q)$. Thus, (b) holds. The proof that (c) is true follows in a similar way and is left as an exercise. This shows that $\mathcal{B}$ is a base and, therefore, generates some topology. The fact that it generates the topology defined by the metric is left for the reader to verify. ∎

The proof of the next corollary is immediate.

Corollary 2.2.5. *Let $\mathcal{S}$ be any collection of subsets of an arbitrary set X such that:*

(a) $X = \bigcup\{S : S \in \mathcal{S}\}$;

(b) *for any pair of distinct points x, y in X there are disjoint sets S, T in $\mathcal{S}$ such that $x \in S$ and $y \in T$.*

If $\mathcal{B}$ consists of all finite intersections of sets from $\mathcal{S}$, then $\mathcal{B}$ is a base for a topology.

This leads to the following definition, which is our second way of generating a topology.

Definition 2.2.6. If $\mathcal{S}$ is any collection of subsets of an arbitrary set X having properties (a) and (b) in the preceding corollary, then $\mathcal{S}$ is called a *subbase*. The collection $\mathcal{B}$ of all finite intersections of sets from $\mathcal{S}$ is called the *base generated by $\mathcal{S}$* (see the preceding corollary). The topology defined by this base is called the *topology generated by $\mathcal{S}$*.

Once again, the topology generated by a subbase is unique (Exercise 2). The virtue of a subbase is that it is so easy to find a collection of sets that satisfies the definition. This facilitates inventing examples, though it might make it more difficult to verify that a topology generated by a subbase has some specific property. Therefore, when we introduce concepts for a topological space, we should try to give an equivalent formulation of the concept in terms of a subbase as well as a base.

Example 2.2.7.   (a) Every base for a topology is a subbase.

(b) If $X = \mathbb{R}$ and $\mathcal{S}$ consists of all the intervals of the form $(-\infty, a)$ and (a, ∞), then $\mathcal{S}$ is a subbase for the usual topology on $\mathbb{R}$. Of course, we can restrict the numbers a in these intervals to be rational or irrational, and we still have a subbase that generates the usual topology. Observe that $\mathcal{S}$ is not a base for the topology of $\mathbb{R}$.

Exercises

(1) Show that the topology generated by a base $\mathcal{B}$ is the intersection of all topologies that contain $\mathcal{B}$.

(2) Prove that the topology generated by a subbase is the intersection of all topologies that contain it.

(3) For a topological space $(X, \mathcal{T})$ give a necessary and sufficient condition on a collection of subsets $\mathcal{C}$ of X that $\mathcal{B} = \{X \backslash C : C \in \mathcal{C}\}$ is a base for the topology of X. Can we say that such a collection $\mathcal{C}$ is a base for the closed subsets of $(X, \mathcal{T})$?

(4) Consider the plane $\mathbb{R}^2$, and define an open half-plane to be a set of the form $\{(x, y) \in \mathbb{R}^2 : ax + by < c\}$ for some choice of constants a, b, c. (a) If $a, b, c \in \mathbb{R}$, show that $\{(x, y) \in \mathbb{R}^2 : ax + by > c\}$ is an open half-plane. (b) Show that the collection of all open half-planes is a subbase for the usual topology on $\mathbb{R}^2$.

(5) A *partially ordered set* is a pair $(X, \leq)$, where X is a set and $\leq$ is a relation on X such that: (i) $x \leq x$ for all X; (ii) if $x \leq y$ and $y \leq z$, then $x \leq z$; (iii) if $x \leq y$ and $y \leq x$, then $x = y$. (Also see Definition A.4.1.) Say that a partially ordered set is *linearly ordered* if whenever x and y are in X, either $x \leq y$ or $y \leq x$. (For example, $X = \mathbb{R}$ or any of its subsets is linearly ordered.) (a) If X is a partially ordered set and $\mathcal{S}$ is the collection of all sets having the form of either $\{y : y \leq x \text{ and } y \neq x\}$ or $\{y : x \leq y \text{ and } y \neq x\}$, show that $\mathcal{S}$ is a subbase for a topology

on X, though it may not satisfy the Hausdorff property. This is called
the *order topology* on X. (b) Show that if $(X, \leq)$ is linearly ordered,
then the order topology satisfies the Hausdorff property. Can you find
another condition on the ordering such that the order topology has
the Hausdorff property? (c) When a and b are elements of a partially
ordered space, let $(a, b) = \{x \in X : a < x < b\}$. If $(X, \leq)$ is linearly
ordered, show that $\{(a, b) : a, b \in X \text{ and } a < b\}$ is a base of the order
topology.

(6) This exercise is taken from [1]. Define the space ℓ^1 of absolutely sum-
mable sequences: $\ell^1 = \{\{a_n\} : a_n \in \mathbb{R} \text{ and } \sum_{n=1}^{\infty} |a_n| < \infty\}$. (a)
Show that $\rho(a, b) = \sum_{n=1}^{\infty} |a_n - b_n|$ defines a metric on ℓ^1. (b) Show
that if $u = \{a_n\} \in \ell^1$ and $x = \{x_n\} \in \ell^\infty$ (Exercise 1.1.12), then
$\sum_{n=1}^{\infty} |a_n x_n| < \infty$. We use the notation $\langle a, x \rangle = \sum_{n=1}^{\infty} a_n x_n$. When
$a \in \ell^1$, $x \in \ell^\infty$, and $\epsilon > 0$, let

$$U_{x,\epsilon}(a) = \{b \in \ell^1 : |\langle a - b, x \rangle| < \epsilon\}.$$

(c) Show that $\mathcal{S} = \{U_{x,\epsilon}(a) : a \in \ell^1, x \in \ell^\infty, \epsilon > 0\}$ is a subbase for
a topology on ℓ^1. Let $\mathcal{T}$ be the topology on ℓ^1 defined by the subbase
$\mathcal{S}$. We will show that the topology $\mathcal{T}$ differs from that defined by the
metric ρ on ℓ^1. (d) If $S = \{a \in \ell^1 : \sum_{n=1}^{\infty} |a_n| = 1\}$, show that the
closure of S in the $\mathcal{T}$ topology includes the sequence $0 = \{0, 0, \dots\}$.
That is, for $x_1, \dots, x_m \in \ell^\infty$ and positive $\epsilon_1, \dots, \epsilon_m$, there is an a in S
with $a \in \bigcap_{k=1}^{m} U_{x_k, \epsilon_k}(0)$. Note that this says that 0 is a limit point of S.
(In fact, the closure of S in the topology $\mathcal{T}$ can be shown to be $\{a \in \ell^1 :
\sum_{n=1}^{\infty} |a_n| \leq 1\}$.) (e) Consider $X = \{x \in \ell^\infty : d(x, 0) = \sup_n |x_n| \leq 1\}$;
since ℓ^∞ with its metric d is a complete metric space (Exercise 1.2.11)
and X is a closed subset of ℓ^∞, we note that (X, d) is a complete metric
space. (f) For each x in X, $\delta > 0$, and positive integer N, define

$$G(x; \delta, N) = \{y \in X : |x_n - y_n| < \delta \text{ for } 1 \leq n \leq N\}.$$

Show that $\mathcal{B} = \{G(x; \delta, J) : x \in X, \delta > 0, N \geq 1\}$ is a base for the
topology on (X, d). (g) If $a \in \ell^1$, show that $x \mapsto \langle a, x \rangle$ is a continuous
function on (X, d). (h) Now assume that $\{a^k\}$ is a sequence in ℓ^1 with
$a^k = \{a_1^k, a_2^k, \dots\}$. Show that if $a^k \to 0$ in $(\ell^1, \mathcal{T})$ (recall the definition of
a convergent sequence in a topological space given just after Definition
2.1.10), then for every $m \geq 1$ and any $\epsilon > 0$

$$F_m = \{x \in X : |\langle a^k, x \rangle| \leq \epsilon/3 \text{ for } k \geq m\}$$

is a closed subset of (X, d). Maintain this notation for the rest of the
exercise. (i) Use the Baire Category Theorem to show that there is an
integer m, a point x in X, a $\delta > 0$, and a positive integer N such that
$G(x; \delta, N) \subseteq F_m$. (j) Conclude that $\rho(a_k, 0) \to 0$. (k) Use part (d)
to conclude that the topology $\mathcal{T}$ on ℓ^1 is different from the topology
defined by the metric ρ.

2.3. Continuous Functions

Here we extend to the setting of topological spaces the definition of a continuous function given in §1.3 for metric spaces. The original definition in the metric space environment (Definition 1.3.1) involves the metric, so that will not extend. We could define convergent sequences in a topological space and use that, but it is inadequate for reasons we cannot go into here. Nevertheless, we can use Theorem 1.3.3, which gives an equivalent formulation of continuity in terms of open sets.

Definition 2.3.1. If $(X, \mathcal{T})$ and $(W, \mathcal{U})$ are topological spaces and $f : X \to W$, say that f is *continuous* at a point x if for every neighborhood U of $f(x)$ in W, there is a neighborhood G of x in X such that $f(G) \subseteq U$ (equivalently, $G \subseteq f^{-1}(U)$). Say that f is continuous on X if it is continuous at each point.

As we said in discussing continuous functions on a metric space, we will not focus on functions continuous at a point, but we will put all our energy into understanding functions that are continuous on the entire topological space.

Proposition 2.3.2. *If $f : (X, \mathcal{T}) \to (W, \mathcal{U})$, then the following statements are equivalent:*

(a) *f is continuous.*

(b) *For every open set U in W, $f^{-1}(U)$ is open in X.*

(c) *For every closed set C in W, $f^{-1}(C)$ is closed in X.*

(d) *If $\mathcal{B}$ is a base for the topology of W, then $f^{-1}(B) \in \mathcal{T}$ for every A in $\mathcal{B}$.*

(e) *If $\mathcal{S}$ is a subbase for the topology of W, then $f^{-1}(S) \in \mathcal{T}$ for every S in $\mathcal{S}$.*

Proof. It is left to the reader to show that (a), (b), and (c) are equivalent. (See the proof of Theorem 1.3.3, which can essentially be lifted to this setting.) Since a base is contained in the topology, (b) implies (d); since a subbase is contained in a base, (d) implies (e). If (e) holds, observe that $\{U \in \mathcal{U} : f^{-1}(U) \in \mathcal{T}\}$ is closed under finite intersections and arbitrary unions and contains $\mathcal{S}$; hence it equals $\mathcal{U}$. This implies that (b) holds and completes the proof of the proposition. ∎

Since the open balls in a metric space form a basis for the topology, the equivalence of (a) and (c) can be considered the extension of the $\epsilon - \delta$ definition of continuity given for a function from one metric space into another.

Proposition 2.3.3. *Let $(X, \mathcal{T}), (W, \mathcal{U})$, and $(Y, \mathcal{V})$ be topological spaces. If $f : (X, \mathcal{T}) \to (Y, \mathcal{V})$ and $g : (Y, \mathcal{V}) \to (W, \mathcal{U})$ are continuous functions, then so is the composition $g \circ f : (X, \mathcal{T}) \to (W, \mathcal{U})$.*

Proof. If $U \in \mathcal{U}$, then $(g \circ f)^{-1}(U) = f^{-1}\left[g^{-1}(U)\right]$. But $g^{-1}(U) \in \mathcal{V}$, so that $(g \circ f)^{-1}(U) \in \mathcal{T}$, whence the continuity of $g \circ f$. ∎

Even though the next result is almost longer to state than it is to prove, it is a useful way to put together continuous functions to form another.

Proposition 2.3.4. *Let $(X, \mathcal{T})$ and $(W, \mathcal{U})$ be topological spaces with subsets A and B such that $X = A \cup B$, suppose $g : A \to W$ and $h : B \to W$ are continuous functions such that $g(x) = h(x)$ when $x \in A \cap B$, and define $f : X \to W$ by letting $f(x) = g(x)$ when $x \in A$ and $f(x) = h(x)$ when $x \in B$. If both A and B are open or if both are closed, then the function f is continuous.*

Proof. Assume both A and B are open. If U is an open subset of W, then it is easily verified that $f^{-1}(U) = g^{-1}(U) \cup h^{-1}(U)$. Now $g^{-1}(U)$ and $h^{-1}(U)$ are open subsets of A and B, respectively, and, since A and B are open in X, $f^{-1}(U)$ is open. Therefore, f is continuous. The proof where the sets A and B are closed is similar. ∎

In conjunction with the last proposition, see Exercises 2 and 3.

Now we want to define the cartesian product of a finite number of topological spaces, similar to our definition (§ 1.1) of the product of a finite number of metric spaces. Here, however, there is a more "canonical" approach to the topology than there was in the metric space setting. Recall that when we defined the product of metric spaces, we chose one metric on the product space. Later we saw some equivalent metrics. Of course, when metrics are equivalent, they generate the same topology. That is the key to this more general setting, though here there is no middleman.

Proposition 2.3.5. *If $(X_k, \mathcal{T}_k)$ is a topological space for $1 \le k \le n$ and $X = X_1 \times \cdots \times X_n$, then $\mathcal{B} = \{G_1 \times \cdots \times G_n : G_k \in \mathcal{T}_k \text{ for } 1 \le k \le n\}$ is a base for a topology on X. If $\mathcal{T}$ is the topology defined by $\mathcal{B}$, then $\mathcal{T}$ is the smallest of all the topologies $\mathcal{U}$ on X such that for $1 \le k \le n$ the projections $(X, \mathcal{U}) \to (X_k, \mathcal{T}_k)$ defined by $(x_1, \ldots, x_n) \mapsto x_k$ are continuous.*

Proof. Clearly, $\mathcal{B}$ covers X. Suppose $(x_1, \ldots, x_n) = x \ne y = (y_1, \ldots, y_n)$. So for some integer k, $x_k \ne y_k$, and there must be disjoint sets G_k, H_k belonging to $\mathcal{T}_k$ with x_k in G_k and y_k in H_k. For notational convenience, assume $k = 1$. If $G'_1 = G_1 \times X \times \cdots \times X$ and $H'_1 = H_1 \times X \times \cdots \times X$, then $x \in G'_1$, $y \in H'_1$, and $G'_1 \cap H'_1 = \emptyset$. So $\mathcal{B}$ has the Hausdorff property. Now let $x = (x_1, \ldots, x_n) \in G \cap H$, where $G = G_1 \times \cdots \times G_n$, $H = H_1 \times \cdots \times H_n$, and each G_k, H_k belongs to $\mathcal{T}_k$. If $U = (G_1 \cap H_1) \times \cdots \times (G_n \cap H_n)$, then $U \in \mathcal{B}$ and $x \in U \subseteq G \cap H$. Therefore $\mathcal{B}$ is a base.

Let $\pi_1 : X \to X_1$ be the map defined by $\pi_1(x_1, \ldots, x_n) = x_1$. Note that if $U_1 \in \mathcal{T}_1$, then $\pi_1^{-1}(U_1) = U_1 \times X_2 \times \cdots \times X_n$. Thus, π_1 is continuous; similarly, each π_k is continuous. Let $\mathcal{U}$ be another topology on X such that $\pi_k : (X, \mathcal{U}) \to X_k$ is continuous for $1 \le k \le n$. If $G_k \in \mathcal{T}_k$ for $1 \le k \le n$, then we have that $G_1 \times \cdots \times G_n = (G_1 \times X_2 \times \cdots \times X_n) \cap \cdots \cap (X_1 \times X_2 \times \cdots \times X_{n-1} \times G_n) = \pi_1^{-1}(G_1) \cap \cdots \pi_n^{-1}(G_n) \in \mathcal{U}$. Thus, $\mathcal{B} \subseteq \mathcal{U}$, and it follows that $\mathcal{T} \subseteq \mathcal{U}$. ∎

Definition 2.3.6. With the notation as in the preceding proposition, the topology $\mathcal{T}$ is called the *product topology* on X and the maps $(x_1, \ldots, x_n) \mapsto x_k$ are called the *coordinate projections*, or simply *projections*. Usually each projection map will be denoted by π_k, as in the preceding proof.

It is left to the reader to verify that the product topology that was just defined agrees with the product topology defined previously on the product of metric spaces (Exercise 5). Later ($\S 2.6$) we will define the product topology on an infinite product of topological spaces, but first we will explore finite products. We want to establish an important property of the projection maps, but this requires another definition. To facilitate this discussion and later ones, we are going to start an economy of expression by not necessarily specifying the name of the topology on a topological space. So instead of saying, "Let $(X, \mathcal{T})$ be a topological space," we will just say, "Let X be a topological space." There will seldom be a problem, but we can always return to the old phrasing if there is some ambiguity or a need for specificity.

Definition 2.3.7. If X and Z are topological spaces and $f : X \to Z$, then f is an *open map* provided $f(G)$ is open in Z whenever G is open in X.

Proposition 2.3.8. *If $X_1, \ldots, X_n$ are topological spaces and $X = X_1 \times \cdots \times X_n$ has the product topology, then each projection map is an open map.*

Proof. Let G be an open subset of X, let $1 \le k \le n$, and suppose $x_k \in \pi_k(G)$; to show that $\pi_k(G)$ is open, we need to show that there is an open subset G_k of X_k such that $x_k \in G_k \subseteq \pi_k(G)$. Now there is an x in G with $\pi_k(x) = x_k$; that is, $x = (x_1, \ldots, x_n)$, and the kth coordinate is precisely the same x_k we started with. By Proposition 2.3.5, there are open sets G_j in X_j, $1 \le j \le n$, with $x = (x_1, \ldots, x_n) \in G_1 \times \cdots \times G_n \subseteq G$. It follows that $x_k \in G_k = \pi_k(G_1 \times \cdots \times G_n) \subseteq \pi_k(G)$. ∎

Proposition 2.3.9. *If $X_1, \ldots, X_n$ are topological spaces, $X = X_1 \times \cdots \times X_n$ has the product topology, Y is another topological space, and $f : Y \to X$, then f is continuous if and only if $\pi_k \circ f : Y \to X_k$ is continuous for $1 \le k \le n$.*

Proof. If f is continuous, then $\pi_k \circ f$ is the composition of two continuous functions, and this is continuous by Proposition 2.3.4. Now assume that $\pi_k \circ f$ is continuous for $1 \le k \le n$. If G_k is open in X_k for each k, then $f^{-1}(G_1 \times \cdots \times G_n) = \{y \in Y : f(y) \in G_1 \times \cdots \times G_n\} = \{y \in Y : \pi_k(f(y)) \in G_k \text{ for } 1 \le k \le n\} = \bigcap_{k=1}^{n} (\pi_k \circ f)^{-1}(G_k)$. Since each $\pi_k \circ f$ is continuous, each $(\pi_k \circ f)^{-1}(G_k)$ is open, and so $f^{-1}(G_1 \times \cdots \times G_n)$ is open. By Proposition 2.3.2, f is continuous. ∎

The next proposition could have been presented earlier, but it would have required a different, less efficient proof.

Proposition 2.3.10. *If $f, g : X \to \mathbb{R}$ are continuous functions, then so are $f + g : X \to \mathbb{R}$ and $fg : X \to \mathbb{R}$.*

Proof. First consider the function $s : \mathbb{R} \times \mathbb{R} \to \mathbb{R}$ defined by $s(a, b) = a + b$. We leave the reader to use sequences to show that s is continuous. Now note that $(f + g) : X \to \mathbb{R}$ is the composition of s and the function $\phi : X \to \mathbb{R} \times \mathbb{R}$ defined by $\phi(x) = (f(x), g(x))$. An easy application of the preceding proposition shows that ϕ is continuous. Thus $f + g$ is continuous. The proof that fg is continuous is similar (Exercise 6). ∎

Here are other ways to manufacture continuous functions from old ones, provided the range of the functions is the real numbers. If $f, g : X \to \mathbb{R}$, define

$$(f \vee g)(x) = \max\{f(x), g(x)\},$$
$$(f \wedge g)(x) = \min\{f(x), g(x)\},$$
$$|f|(x) = |f(x)|$$

for all x in X. It is useful to note that $f \wedge g = -[(-f) \vee (-g)]$ and $f \vee g = -[(-f) \wedge (-g)]$, which can be verified by evaluating at a point x in X and using the corresponding equalities for real numbers. Hence many arguments involving one operation can be obtained immediately from the other. We do this in the next proof.

Proposition 2.3.11. *If $f, g : X \to \mathbb{R}$ are continuous functions, then so are $f \vee g$, $f \wedge g$, and $|f|$.*

Proof. That the map $x \mapsto (f(x), g(x))$ is continuous from X into $\mathbb{R}^2$ is immediate from Proposition 2.3.9. We will show that the map $(s, t) \mapsto s \vee t \equiv \max\{s, t\}$ is a continuous map from $\mathbb{R}^2$ into $\mathbb{R}$. Once this is done, we see that $f \vee g$ is the composition of two continuous functions and is thus continuous. We use a sequential argument. Suppose $s_n \to s$ and $t_n \to t$. If $t > s$, then there is an integer N such that $t_n - s_n > 0$ for all $n \geq N$. Thus, for $n \geq N$, $s_n \vee t_n = t_n \to t = s \vee t$. Similarly, when $s > t$, $s_n \vee t_n \to s \vee t$. Assume $s = t$, so $s \vee t = t$ and $\lim(t_n - t) = 0 = \lim_n (s_n - t)$. If $\epsilon > 0$, then there is an $N \geq 1$ such that for $n \geq N$, $t - \epsilon < t_n, s_n < t + \epsilon$, and thus $s \vee t - \epsilon = t - \epsilon < s_n \vee t_n < t + \epsilon = s \vee t + \epsilon$. Hence this map is continuous.

As we observed prior to the statement of this proposition, $f \wedge g = -[(-f) \vee (-g)]$, so it too is continuous. Finally, $|f|$ is the composition of the continuous functions f and $t \mapsto |t|$ defined on $\mathbb{R}$. ∎

Definition 2.3.12. If X and Z are topological spaces, then a *homeomorphism* between X and Z is a bijection $f : X \to Z$ that is continuous and has a continuous inverse $f^{-1} : Z \to X$.

We note that this extends the definition of a homeomorphism between two metric spaces given in §1.3. Also, a bijection f is a homeomorphism if and only if f is both continuous and open. (Why?) As in the situation of metric spaces, it is easy to see that topological spaces being homeomorphic is an equivalence relation (Definition 2.8.1), the basic notion of equivalence among topological spaces.

Exercises

(1) Given the fact that the collection of all open infinite intervals $(-\infty, a)$, (a, ∞) forms a subbase for the topology of $\mathbb{R}$, does the equivalence of continuity and condition (d) in Proposition 2.3.2 remind you of anything for functions $f : \mathbb{R} \to \mathbb{R}$?

(2) In Proposition 2.3.4, construct a counterexample to the conclusion if you do not assume that A and B are both open or both closed.

(3) Suppose $(X, \mathcal{T})$ and $(W, \mathcal{U})$ are topological spaces, $X = \bigcup_i A_i$, for each i there is a continuous function $g_i : A_i \to W$ such that $g_i(x) = g_j(x)$ when $x \in A_i \cap A_j$, and $f : X \to W$ is defined by $f(x) = g_i(x)$ when $x \in A_i$. Can Proposition 2.3.4 be extended to this situation?

(4) If $(X_k, \mathcal{T}_k)$ is a topological space for $1 \le k \le n$ and $X = X_1 \times \cdots \times X_n$, show that $\{\pi_k^{-1}(G_k) : 1 \le k \le n \text{ and } G_k \in \mathcal{T}_k\}$ is a subbase for the product topology on X.

(5) Let (X_k, d_k) be a metric space for $1 \le k \le n$, and let $\mathcal{T}_k$ be the resulting topology. Show that the topology defined on the cartesian product $X = X_1 \times \cdots \times X_n$ by the product metric as in Definition 1.1.16 is the same as the product topology defined on X after Proposition 2.3.5. (Readers may assume that $n = 2$ if they so desire.)

(6) Give the details showing that fg in Proposition 2.3.10 is continuous.

(7) If X_k, Z_k are topological spaces for $1 \le k \le n$, $X = X_1 \times \cdots \times X_n$, and $Z = Z_1 \times \cdots \times Z_n$, show that X and Z are homeomorphic if and only if, after some renumbering of the spaces, X_k and Z_k are homeomorphic for $1 \le k \le n$.

(8) Is an open map also a closed map, that is, a function that maps closed sets into closed sets?

(9) Let $\mathcal{F}$ denote all the functions from a set X into $\mathbb{R}$, and consider the binary operations on $\mathcal{F}$ defined by $\vee : \mathcal{F} \times \mathcal{F} \to \mathcal{F}$ and $\wedge : \mathcal{F} \times \mathcal{F} \to \mathcal{F}$. (a) Are these operations associative? (b) Are they distributive: $f \vee (g \wedge h) = (f \vee g) \wedge (f \vee h)$ or $f \wedge (g \vee h) = (f \wedge g) \vee (f \wedge h)$?

2.4. Compactness and Connectedness

In this section we will revisit two concepts we saw for metric spaces, compactness and connectedness, and we will also introduce an additional form of connectedness. In the following section, we will see a stronger form of connectedness. We start with compactness, whose definition in a topological space is the same as for a metric space.

Definition 2.4.1. A subset K of a topological space X is *compact* if every open cover of K has a finite subcover.

In this section, when we prove Theorem 2.4.6 on compactness below, we will use Zorn's Lemma (Theorem A.4.6). If you are not familiar with this result, you must study § A.4, and I advise doing so before you start the present section. This material will be used in the future as we progress, and no notice such as this will be given again.

We start with some basic properties of compactness. It is clear that a subset is compact if and only if it is compact as a topological space when it is given its relative topology, that is, if and only if every cover by relatively open sets has a finite subcover. The proof of the first proposition (Exercise 1) is an imitation of the proof of Proposition 1.4.2 as is its corollary, but where open sets replace open balls. Note that there is no concept of a bounded set in an abstract topological space.

Proposition 2.4.2. *Let X be a topological space, and let $K \subseteq X$.*

(a) *If K is a compact subset of X, then K is closed.*

(b) *If K is compact and F is a closed set contained in K, then F is compact.*

(c) *The continuous image of a compact set is compact.*

Corollary 2.4.3. *If X is a compact space and $f : X \to \mathbb{R}$ is a continuous function, then there are points a and b in X such that $f(a) \le f(x) \le f(b)$ for all x in X.*

We will use the concept of a collection of subsets of a topological space having the finite intersection property (FIP). Because the definition is the same for a topological space as for a metric space, we will not repeat it.

Proposition 2.4.4. *If K is a closed subset of a topological space X, then K is compact if and only if every collection of closed subsets of K having the FIP has a nonempty intersection.*

The proof is exactly the same as the proof of the equivalence of (a) and (b) in Theorem 1.4.5. The other parts of Theorem 1.4.5 are not true in a nonmetric space; indeed, some, such as the condition involving total boundedness, do not even make sense. Also, in an abstract topological space, the idea of completeness is nonsensical since we cannot define a Cauchy sequence. Later, when we define the extension of the concept of a sequence, we will see the natural extension of the equivalence with compactness of the parts of Theorem 1.4.5 that involve sequences.

Proposition 2.4.5. *If $\mathcal{B}$ is a base for the topology for X and $K \subseteq X$, then K is compact if and only if every cover of K by sets from $\mathcal{B}$ has a finite subcover.*

Proof. Since $\mathcal{B}$ is contained in the collection of open sets, the stated condition is easily seen to follow from the assumption that K is compact. Now assume that every cover of K by sets from the base has a finite subcover, and let $\mathcal{C}$ be an open cover of K. If $G \in \mathcal{C}$, then there is a subset $\mathcal{B}_G$ of the base $\mathcal{B}$ such that $G = \bigcup\{B : B \in \mathcal{B}_G\}$. Thus, $\{B : B \in \mathcal{B}_G$ and $G \in \mathcal{C}\}$ is a cover of K by sets from the base. By assumption, there is a finite subcover; that is, there are $G_1, \ldots, G_n$ in $\mathcal{C}$ and sets $B_1, \ldots, B_n$ in $\mathcal{B}$ such that $B_k \subseteq G_k$ for

$1 \leq k \leq n$ and $K \subseteq \bigcup_{k=1}^{n} B_k$. Since $B_k \subseteq G_k$, $\{G_1, \ldots, G_n\}$ constitutes the sought-after finite subcover of $\mathcal{C}$.   ∎

Here is where we make use of Zorn's Lemma. The proof of the next theorem will deal with collections of collections of subsets of a topological space X. That is, it will treat sets Γ that are contained in 2^{2^X}, so stay on your toes. To help out, we will reserve capital Greek letters like Γ for subsets of 2^{2^X}, capital script letters like $\mathcal{W}$ for subsets of 2^X or elements of Γ, and capital roman letters like W for subsets of X. So $W \in \mathcal{W} \in \Gamma$.

Theorem 2.4.6 (Alexander's[2] Theorem). *If X is a topological space and $\mathcal{S}$ is a subbase for the topology of X, then X is compact if and only if every cover by sets from $\mathcal{S}$ has a finite subcover.*

Proof. Again, as in the preceding proof, it is clear that if X is compact, then every subbasic cover of X has a finite subcover. To prove the converse, assume it is false: that is, assume that every cover by sets from $\mathcal{S}$ has a finite subcover, but X is not compact. Let Γ denote the collection of all open covers of X that do not have a finite subcover; since X is not compact, $\Gamma \neq \emptyset$. Order Γ by inclusion (§ A.4). It is easy to see that if Λ is a chain in Γ, then $\mathcal{V} = \bigcup \{ \mathcal{W} : \mathcal{W} \in \Lambda \}$ is an open cover of X. If $\mathcal{V}$ has a finite subcover $G_1, \ldots, G_n$, then for $1 \leq k \leq n$ there is a $\mathcal{W}_k$ in Γ such that $G_k \in \mathcal{W}_k$. But since Λ is a chain, one of the covers $\mathcal{W}_k$ contains all the others. That is, all the sets $G_1, \ldots, G_n$ belong to one of the covers $\mathcal{W}_k$; thus, this $\mathcal{W}_k$ has a finite subcover, contradicting the fact that it belongs to Γ. Thus, $\mathcal{V} \in \Gamma$, so that it is an upper bound of the arbitrary chain Λ. By Zorn's Lemma (Theorem A.4.6), Γ has a maximal element $\mathcal{C}$—an open cover of X without a finite subcover. Note that if G is any nonempty open subset of X and $G \notin \mathcal{C}$, then $\mathcal{C} \cup \{G\}$ is an open cover of X that is strictly larger than $\mathcal{C}$. Thus $\mathcal{C} \cup \{G\}$ has a finite subcover by the maximality of $\mathcal{C}$ in Γ; that is, $\mathcal{C} \cup \{G\} \notin \Gamma$. (Observe that the finite subcover of $\mathcal{C} \cup G$ must include the set G.)

Let $\mathcal{W} = \mathcal{C} \cap \mathcal{S}$; that is, $\mathcal{W}$ consists of all the subbasic sets belonging to the cover $\mathcal{C}$. (Yes, $\mathcal{W}$ is possibly empty.) $\mathcal{W}$ cannot cover X since then, by

---

[2]James Waddell Alexander was born in 1888 in Sea Bright, New Jersey. His father was the American painter John White Alexander. In 1915 he received his doctorate from Princeton University, having previously spent time studying mathematics in Bologna and Paris. He married in 1917. During World War I he entered the army as a lieutenant and at the end of the war left as a captain. He returned to Princeton as an Assistant Professor in 1920 and was promoted in 1928 to Professor. From 1933 on he was a member of the Institute for Advanced Study. During World War II he was a civilian working with the U.S. Army Air Force at their Office of Scientific Research and Development. Because of his leftist political views he came under the scrutiny of the McCarthy Committee; this had the effect of turning him into a recluse after his retirement in 1951. His research focused on topology, particularly algebraic topology, in which he was a pioneer of cohomology theory. His named contributions include the Alexander Duality Theorem, the Alexander horned sphere, the present result, and the Alexander polynomial used in knot theory. There is also Alexander–Spanier cohomology theory, which he introduced in 1935 and that was generalized to its present form by Spanier in 1948. In addition, he impressed those who knew him as a charming man with a fondness for limericks and mountain climbing. His climbing was centered in the Swiss Alps and Colorado Rockies. Alexander's Chimney, in the Rocky Mountain National Park, is named after him. He died in 1971 in Princeton, New Jersey.

hypothesis, $\mathcal{W}$ would have a finite subcover, and thus so would $\mathcal{C}$. Let

2.4.7 $$x \in X \backslash \bigcup_{W \in \mathcal{W}} W,$$

and let $C \in \mathcal{C}$ such that $x \in C$. Since $\mathcal{S}$ is a subbase, there are $S_1, \ldots, S_n$ in $\mathcal{S}$ such that $x \in \bigcap_{k=1}^n S_k \subseteq C$. Note that by (2.4.7), none of the subbasic sets S_k can belong to the cover $\mathcal{C}$. Hence for each $k, 1 \leq k \leq n, \mathcal{C} \cup \{S_k\}$ is a cover of X with a finite subcover. Let $H_1^k, \ldots, H_{m_k}^k \in \mathcal{C}$ be such that

$$X = S_k \cup \bigcup_{j=1}^{m_k} H_j^k.$$

Therefore,

$$X = \bigcap_{r-1}^n \left(S_r \cup \left[\bigcup_{k=1}^n \bigcup_{j=1}^{m_k} H_j^k \right] \right)$$

$$\subseteq \left(\bigcap_{r=1}^n S_r \right) \cup \left[\bigcup_{k=1}^n \bigcup_{j-1}^{m_k} H_j^k \right]$$

$$\subseteq C \cup \bigcup_{k=1}^n \bigcup_{j=1}^{m_k} H_j^k.$$

But this says that $\{C\} \cup \{H_j^k : 1 \leq j \leq m_k, 1 \leq k \leq n\}$ is a finite subcover of $\mathcal{C}$, furnishing the sought-after contradiction. ∎

We will use Alexander's Theorem when we investigate the infinite product of topological spaces in § 2.6. Readers might look at Exercise 4, where they are asked to use Alexander's Theorem to show that the product of a finite number of compact spaces is compact. Now we turn our attention to connectedness.

Definition 2.4.8. A topological space X is *connected* if there are no subsets of X that are both open and closed other than X and the empty set. A subset of X is connected if it is a connected topological space when it has its relative topology.

This extends the definition of connectedness from metric spaces to topological spaces, and so all the examples we saw in § 1.5 remain valid. In particular, we still have that X is connected if and only if whenever $X = A \cup B$ with $A \cap B = \emptyset$ and both A and B open (or closed), then either A or B is the empty set.

Theorem 2.4.9. *The continuous image of a connected set is connected.*

The proof of Theorem 1.5.3 applies almost verbatim in a proof of the preceding result. Similarly, an inspection of the proof of Proposition 1.5.6

reveals that only the properties of open and closed sets were used with no reference to a metric, so the same proof works in the present setting. We will state this result for topological spaces, and neither its proof nor those of other extensions will be given.

Proposition 2.4.10. *Let X be a topological space.*

(a) *If $\{E_i : i \in I\}$ is a collection of connected subsets of X such that $E_i \cap E_j \neq \emptyset$ for all i, j in I, then $E = \bigcup_{i \in I} E_i$ is connected.*

(b) *If $\{E_n : n \geq 1\}$ is a sequence of connected subsets of X such that $E_n \cap E_{n+1} \neq \emptyset$ for each n, then $E = \bigcup_{n=1}^{\infty} E_n$ is connected.*

Corollary 2.4.11. *The union of two intersecting connected subsets of a topological space is connected.*

As before, we define a *component* of X as a maximal connected subset of X.

Proposition 2.4.12. *For any topological space every connected set is contained in a component, distinct components are disjoint, and the union of all the components is the entire space.*

Proposition 2.4.13. *If C is a connected subset of the topological space X and $C \subseteq Y \subseteq \operatorname{cl} C$, then Y is connected.*

Corollary 2.4.14. *The closure of a connected set is connected, and each component is closed.*

We end this section with a variation on the idea of connectedness.

Definition 2.4.15. A topological space X is *locally connected* if for each x in X and every neighborhood G of x there is a neighborhood U of x such that $U \subseteq G$ and U is connected.

This is the first example we have encountered of a "local" property, that is, a property that may not be possessed by the entire topological space but holds in arbitrarily small neighborhoods. (This phrase, "in arbitrarily small neighborhoods," is justified by the part of the definition that says there is a connected neighborhood of x inside any given neighborhood G.) We will encounter this again in the next section as well as in §3.5 below when we study the idea of local compactness.

Example 2.4.16.   (a) A connected space is not necessarily locally connected. For example, the topologist sine curve (Example 1.5.12) is connected but not locally connected. (Why?)

(b) A locally connected space is not necessarily connected. For example, a discrete topological space that has more than a single point is locally connected, as is the union of two disjoint closed intervals in $\mathbb{R}$. The reader should have no trouble discovering many more such examples.

(c) Any open subset of $\mathbb{R}^q$ is locally connected.

Proposition 2.4.17. *A topological space X is locally connected if and only if it has a base for its topology consisting of connected sets.*

Proof. If X is locally connected and $\mathcal{B}$ is the collection of all open connected subsets of X, then the definition of local connectedness shows that $\mathcal{B}$ is a base for the topology of X. The proof of the converse is left to the reader. ∎

Proposition 2.4.18. *If X is locally connected, then every component is both open and closed.*

Proof. In general, components are closed. Now assume X is locally connected and C is a component. If $x \in C$, then there is a neighborhood G of x that is connected. But $C \cap G \neq \emptyset$, so $C \cup G$ is connected. Since C is a component, this implies $G \subseteq C$. Since x was an arbitrary point in C, this shows that C is open. ∎

Proposition 2.4.19. *If X and Z are topological spaces with X locally connected and $f : X \to Z$ is continuous, open, and surjective, then Z is locally connected.*

Proof. Let $z \in Z$, and let G be any neighborhood of z; pick any point x in X with $f(x) = z$. By hypothesis, $f^{-1}(G)$ is a neighborhood of x. Since X is locally connected, there is a connected neighborhood V of x such that $V \subseteq f^{-1}(G)$. Since f is continuous and open, $f(V)$ is a connected neighborhood of z, and it is contained in G. By definition, Z is locally connected. ∎

Exercises

(1) Give the details of the proof of Proposition 2.4.2.
(2) Show that the union of a finite number of compact sets is compact.
(3) If X and Z are compact topological spaces and $f : X \to Z$ is a continuous bijection, show that f is a homeomorphism.
(4) Use Alexander's Theorem (2.4.6) to prove that the finite product of compact spaces is compact.
(5) Give the set $\{0, 1\}$ the discrete topology. Show that X is connected if and only if every continuous function from X into the set $\{0, 1\}$ is a constant function.
(6) Give an example of two connected subsets of a topological space whose intersection is not connected.
(7) Show that the product of a finite number of topological spaces $X_1, \ldots, X_n$ is locally connected if and only if each X_k is locally connected.

2.5. Pathwise Connectedness

In this section, we study a further refinement of connectedness, but first we need an additional concept.

Definition 2.5.1. If X is a topological space and $p, q \in X$, a *path* in X from p to q is a continuous function $f : [0, 1] \to X$ such that $f(0) = p$ and $f(1) = q$. The point p is called the *initial point*, and q is the *final point*. If $p = q$, then f is called a *closed path* or *loop*.

It should be noted that what we call a path is called an *arc* by some authors.

The reader has already seen many examples of paths in calculus, and some of the examples below are undoubtedly familiar. But first let us note that the use of the unit interval $[0, 1]$ is just a convenience; any interval would work. In fact, if $a, b \in \mathbb{R}$, $a < b$, and $\tau : [0, 1] \to [a, b]$ is defined by $\tau(t) = tb + (1 - t)a$, then τ is an order-preserving homeomorphism. So if $f : [a, b] \to X$ is continuous, then $f \circ \tau$ is a path as defined previously. Conversely, if $g : [0, 1] \to X$ is a path, then $f = g \circ \tau^{-1} : [a, b] \to X$ is continuous and defines a path in this extended sense. We also note that no matter which interval in $\mathbb{R}$ we use to define a path, the initial and final points are the same. Hence we can define a path in X by giving a continuous function from some other interval than $[0, 1]$. This will frequently be done for convenience. Note that the direction of the path is important, and this is reflected in the use of the order-preserving map τ when we discussed changing the interval used to define a path.

Example 2.5.2. (a) If $x, y \in \mathbb{R}^q$, then $f(t) = ty + (1 - t)x$ is a path in $\mathbb{R}^q$ from x to y. In fact, it traces out the straight line segment from x to y. We abbreviate this by $[x, y]$; note that $[x, y] \neq [y, x]$ since, unless $x = y$, they have different initial and final points.

(b) If $f : [0, 2\pi] \to \mathbb{R}^2$ is given by $f(t) = (\cos t, \sin t)$, then this defines the path that traces out the circle in the plane starting and ending at $(1, 0)$ and traveling in the counterclockwise direction.

(c) If $g : [0, 1] \to X$ is any continuous function, then $f : [0, 1] \to \mathbb{R} \times X$ defined by $f(t) = (t, g(t))$ is a path. In fact, this path traces out the graph of the function g.

Proposition 2.5.3. *If a, b, c are points in a topological space X, f is a path in X from a to b, and g is a path in X from b to c, then*

$$
h(t) = \begin{cases} f(2t) & \text{when } 0 \leq t \leq \frac{1}{2} \\ g(2t - 1) & \text{when } \frac{1}{2} \leq t \leq 1 \end{cases}
$$

is a path in X from a to c.

Proof. An application of Proposition 2.3.4 shows that h is continuous, and you can check that $h(0) = a, h(1) = c$. ∎

The path h defined in the preceding proposition is sometimes called the *product* of f, and g and is denoted by

2.5.4 $h = g \cdot f$

(first follow f and then follow g). Note that the order of these factors is important. Also note that in keeping with the observation that any closed and bounded interval in $\mathbb{R}$ can be used to define a path, we could define

$$g \cdot f(t) = \begin{cases} f(t) & \text{when } 0 \le t \le 1, \\ g(t-1) & \text{when } 1 \le t \le 2. \end{cases}$$

Definition 2.5.5. A topological space X is *pathwise connected* if for any two points p and q in X there is an arc in X from p to q. Another term used for *pathwise connected* is *arcwise connected*.

Needless to say, if E is a subset of a topological space X, then E is said to be pathwise connected if, with its relative topology, it is a pathwise connected topological space. The reader should have no trouble manufacturing examples of pathwise connected proper subsets of $\mathbb{R}^q$.

The reader will see a resemblance between the next proposition and Theorem 2.4.9 and Proposition 2.4.10.

Proposition 2.5.6. *Let X be a topological space.*

(a) *Every pathwise connected space is connected.*

(b) *If X is pathwise connected and Z is another topological space such that there is a continuous surjection $\phi : X \to Z$, then Z is pathwise connected.*

(c) *If $\{E_i : i \in I\}$ is a collection of pathwise connected subsets of X such that $E_i \cap E_j \ne \emptyset$ for all i, j in I, then $E = \bigcup_{i \in I} E_i$ is pathwise connected.*

(d) *If $\{E_n : n \in \mathbb{Z}\}$ is a sequence of pathwise connected subsets of X such that $E_n \cap E_{n+1} \ne \emptyset$ for each n, then $E = \bigcup_{n=1}^{\infty} E_n$ is pathwise connected.*

Proof. (a) Suppose X is not connected, so there are nonempty disjoint open sets A and B with $X = A \cup B$. Pick a in A and b in B. If f is a path in X from a to b, then $f^{-1}(A)$ and $f^{-1}(B)$ are nonempty disjoint open subsets of $[0,1]$ whose union is all of the interval. This contradicts the connectedness of the interval, so that X cannot be pathwise connected.

(b) If p and q are two points in Z, let $a, b \in X$ such that $\phi(a) = p, \phi(b) = q$. If f is a path in X from a to b, then $\phi \circ f$ is a path in Z from p to q.

(c) If $p, q \in E$, let $p \in E_i$ and $q \in E_j$. Fix any point z in $E_i \cap E_j$, let $f : [0,1] \to E_i$ be a path from p to z, and let $g : [0,1] \to E_j$ be a path from z to q. If $h = g \cdot f$, the product of these paths, then it is a path in E from p to q.

(d) If $p, q \in E$, then there are integers m, n such that $p \in E_m$ and $q \in E_n$; assume that $m < n$. By part (c), $E_m \cup E_{m+1}$ is pathwise connected. Again, (c) implies that $E_m \cup E_{m+1} \cup E_{m+2}$ is pathwise connected. Continue and we get that $E_m \cup \cdots \cup E_n$ is pathwise connected. Thus, there is a path in $E_m \cup \cdots \cup E_n \subseteq E$ from p to q. Since p and q were arbitrary, this proves that E is pathwise connected. ∎

Example 2.5.7. Let X be the topologist's sine curve (Example 1.5.12). From (Example 1.5.12) we know that X is connected; however, it is not pathwise connected. See Exercise 4 below as well as Exercise 3.

In light of the preceding example and Proposition 2.5.6(a), pathwise connectedness is a properly stronger property than connectedness. Justified by the preceding proposition, we define a *pathwise connected component* of a topological space as a maximal pathwise connected subset. The next result is the analog of Proposition 2.4.12. The proof is Exercise 5.

Proposition 2.5.8. *For any topological space every pathwise connected set is contained in a pathwise connected component, distinct pathwise connected components are disjoint, and the union of all the pathwise connected components is the entire space.*

It is worth pointing out that Proposition 2.4.13 does not carry over to pathwise connected spaces, as a consideration of the topologist's sine curve with the origin deleted demonstrates.

Now for the local version of pathwise connectivity.

Definition 2.5.9. A topological space X is *locally pathwise connected* if for each x in X and each neighborhood G of x there is another neighborhood U of x that is pathwise connected and such that $U \subseteq G$.

Example 2.5.10. (a) $\mathbb{R}^q$ is locally pathwise connected, as is every discrete space.
   (b) Every open subset of a locally pathwise connected space with its relative topology is also locally pathwise connected. In particular, every open subset of $\mathbb{R}^q$ is locally pathwise connected.
   (c) If X is a metric space, then X is locally pathwise connected if for every x in X and every $\epsilon > 0$ there is a $\delta > 0$ such that $\delta < \epsilon$ and $B(x; \delta)$ is pathwise connected.

It follows that the definition of a locally pathwise connected space can be rephrased as the requirement that the space has a base for its topology consisting of pathwise connected sets.

Proposition 2.5.11. *If X is locally pathwise connected, then an open subset is connected if and only if it is pathwise connected.*

Proof. Let G be an open connected subset of X, and fix a point p in G. If U denotes the pathwise connected component of G that contains p, then the fact that X is locally pathwise connected implies U is an open subset of G. Now let $x \in G$ such that x is in the relative closure of U. If W is a pathwise connected neighborhood of x contained in G, then $W \cap U \neq \emptyset$. Thus, $U \cup W$ is pathwise connected and contained in G. By the definition of a pathwise connected component, $x \in W \subseteq U$. Thus, U is simultaneously open and relatively closed in G. Since G is connected and $U \neq \emptyset$, $U = G$, and so G is pathwise connected. ∎

Corollary 2.5.12. *An open subset of $\mathbb{R}^q$ is connected if and only if it is pathwise connected.*

The proof of the next result is Exercise 8. Also, compare this proposition with Proposition 2.4.19.

Proposition 2.5.13. *If X is locally pathwise connected and $\phi : X \to Z$ is surjective, continuous, and open, then Z is locally pathwise connected.*

Exercises

(1) Giving an example, show that the requirement in Proposition 2.5.6(b) for f to be surjective is necessary.
(2) The following subset of $\mathbb{R}^2$ is often called the *comb*:
$$C = (\{0\} \times [0, 1]) \cup \{(n^{-1}, y) : n \in \mathbb{N}, y \in [0, 1]\} \cup ([0, 1] \times \{0\}).$$
(Why is it called the comb?) Show that the comb is connected. Is it pathwise connected?
(3) If C is the comb (the preceding exercise), then the set $X = C \backslash (\{0\} \times (0, 1))$ is called the *deleted comb*. Show that X is connected but not pathwise connected.
(4) Show that the topologist's sine curve (Example 2.5.7) is not pathwise connected. (Hint: Example 1.3.16(d).)
(5) Prove Proposition 2.5.8.
(6) Verify the statements made in Example 2.5.10.
(7) Give an example of a pathwise connected space that is not locally pathwise connected.
(8) Prove Proposition 2.5.13.
(9) Give an example of a topological space that is not locally pathwise connected but has the property that every point has a neighborhood that is pathwise connected.

2.6. Infinite Products

When we defined the product of two sets X_1, X_2 as the set of all pairs (x_1, x_2) such that $x_j \in X_j$, we could just as easily have defined $X_1 \times X_2$ as the set of functions
$$\{x : \{1, 2\} \to X_1 \cup X_2 : x(j) \in X_j \text{ for } j = 1, 2\}.$$
Similarly, we could redefine a sequence of points in a set X as a function $x : \mathbb{N} \to X$. This may strike you as contrived, but that is because subsets of integers have their natural ordering and we have more experience dealing with n-tuples and sequences than with functions. Now, however, we want to define the product of a set of topological spaces that are indexed by an arbitrary set, not just a subset of the integers. To do this we must adopt the approach using functions. (The reader can also consult § A.2.)

Definition 2.6.1. If I is a nonempty set and for each i in I we have a nonempty set X_i, then the *product* of these sets is defined by

$$\prod_{i \in I} X_i = \left\{ x : I \to \bigcup_{i \in I} X_i : x(i) \in X_i \text{ for all } i \text{ in } I \right\}.$$

We will use the notation x, $\{x_i\}$, or $\{x(i)\}$ for elements of X, depending on the situation and which notation we find convenient. You can do the same.

Do you have trouble making sense of $\bigcup_{i \in I} X_i$? The first time I saw this as a student, it gave me pause since we are not assuming that the sets X_i are all subsets of a common set. If so, think about it for a while, and strictly apply the definition of the union of sets, and I think the trouble will fade. If it does not, use the definition of the disjoint union of sets given in Exercise 1.5.6.

Let us also mention that the Axiom of Choice (§ A.4) is precisely the statement that if $X_i \neq \emptyset$ for all i, then $\prod_{i \in I} X_i \neq \emptyset$.

When each X_i is a topological space, we want to put a topology on the infinite product. To do this, we prove a proposition in a somewhat more general setting. The reader might observe the similarity of the next result with Proposition 2.3.5 for finite products.

Proposition 2.6.2. *Let X be a set, and let $\{X_i : i \in I\}$ be a collection of topological spaces. If, for each i in I, $f_i : X \to X_i$ is a function such that for distinct points x and y in X there is at least one function f_i with $f_i(x) \neq f_i(y)$, then $\mathcal{S} = \{f_i^{-1}(G) : i \in I \text{ and } G \text{ is open in } X_i\}$ is a subbase for a topology $\mathcal{T}$ on X. $\mathcal{T}$ is the smallest of all the topologies $\mathcal{U}$ on X such that $f_i : (X, \mathcal{U}) \to X_i$ is continuous for each i in I.*

Proof. The proof of the first part of this proposition, that $\mathcal{S}$ is a subbase, is straightforward since for any i, $f_i^{-1}(X_i) = X$, and the Hausdorff condition easily follows from the stated assumption about the functions f_i and the fact that each space X_i satisfies the Hausdorff condition. If $\mathcal{T}$ is the topology defined by this subbase $\mathcal{S}$ and $\mathcal{U}$ is another topology such that each $f_i : (X, \mathcal{U}) \to (X_i, \mathcal{T}_i)$ is continuous, then $f_i^{-1}(H) \in \mathcal{U}$ for each H in $\mathcal{T}_i$ and each $i \in I$. But this means that $\mathcal{S} \subseteq \mathcal{U}$, so we have that $\mathcal{T} \subseteq \mathcal{U}$. ∎

The topology $\mathcal{T}$ defined by the collection of functions $\mathcal{F} = \{f_i : i \in I\}$ is called the *weak topology* defined by $\mathcal{F}$. The reader will note the similarity of the next corollary with Proposition 2.3.9.

Corollary 2.6.3. *Adopt the notation of the preceding proposition. If Z is a topological space and $g : Z \to X$ is a function, then g is continuous if and only if $f_i \circ g : Z \to X_i$ is continuous for every i in I.*

Proof. If g is continuous, then $f_i \circ g$ is the composition of two continuous functions and is therefore continuous. Assume each $f_i \circ g$ is continuous, and let $\mathcal{W}$ be the topology of Z. To show that g is continuous, we need only show that $g^{-1}(S) \in \mathcal{W}$ for every S in $\mathcal{S}$. But if $i \in I$ and $G \in \mathcal{T}_i$, then $g^{-1}(f_i^{-1}(G)) = (f_i \circ g)^{-1}(G)$, and this belongs to $\mathcal{W}$ since $f_i \circ g$ is continuous. ∎

If $X = \prod_{i \in I} X_i$, then for each i we can define the projection map $\pi_i : X \to X_i$ by $\pi_i(x) = x(i)$.

Definition 2.6.4. If $\{X_i : i \in I\}$ is a collection of topological spaces, then the weak topology defined on their product X by the projection maps $\{\pi_i : i \in I\}$ is called the *product topology* on X. The subbase $\mathcal{S}$ that appears in Proposition 2.6.2 is called the *standard subbase* for the product topology.

Since the subbase only restricts one coordinate, neighborhoods of a point in the product topology only restrict a finite number of coordinates. It might seem strange that a topology is placed on the product space X by only restricting a finite number of coordinates. Indeed, this was the reaction of some famous topologists at the time this idea was introduced and explored by Andrei Tikhonov in 1926. (See the biographical footnote for Theorem 2.6.7 below.) Nevertheless, this is the most advantageous definition of the product topology, the one universally used. The topology where there are no restrictions on the number of coordinate spaces is called the *box topology*, and some examples are explored in Exercise 3.

We are looking at product spaces, so why did we state and prove Proposition 2.6.2 in a more general setting? Why not proceed with the product and projection maps? The answer is that defining a topology in terms of a collection of functions happens repeatedly in mathematics. In this connection, we call attention to Exercise 2. Also, we will see this subsequently in § 3.4.

The proof of part (a) of the next proposition is like that of Proposition 2.3.8, and part (b) is immediate from Corollary 2.6.3.

Proposition 2.6.5. *Let $\{X_i : i \in I\}$ be a collection of topological spaces and give their product X the product topology.*

(a) *The projection π_i onto X_i is an open map.*

(b) *If Z is a topological space and $g : Z \to X$, then g is continuous if and only if each $\pi_i \circ g : Z \to X_i$ is continuous for each i in I.*

Now we do an about-face from this level of generality and return to metric spaces and use only a countable number of coordinate spaces.

Theorem 2.6.6. *If (X_n, d_n) is a metric space for each $n \geq 1$ and $X = \prod_{n=1}^{\infty} X_n$, then the product topology on X is metrizable and defined by the metric*

$$d(x,y) = \sum_{n=1}^{\infty} \frac{1}{2^n} \frac{d_n(x_n, y_n)}{1 + d_n(x_n, y_n)}.$$

Proof. Using Proposition 1.3.13 we see that the formula for $d(x,y)$ does indeed define a metric on X. Let $\mathcal{D}$ denote the topology defined by the metric d, and let $\mathcal{T}$ be the product topology on X; we must show that $\mathcal{T} = \mathcal{D}$.

If $G \in \mathcal{T}$, then the definition of the product topology implies that for any $x = \{x_n\}$ in G there are $n_1 < \cdots < n_N$ and $\epsilon_1, \ldots, \epsilon_N > 0$ with

$$\bigcap_{k=1}^{N} \pi_{n_k}^{-1}(B(x_{n_k}; \epsilon_k) \subseteq G.$$

Choose $0 < \epsilon < 1$ such that when $0 \le s < 2^N \epsilon$, then $s(1-s)^{-1} < \epsilon_n$ for $1 \le n \le N$. Thus, when $d(x,y) < \epsilon$, $d_n(x_n, y_n)[1 + d_n(x_n, y_n)]^{-1} < 2^N \epsilon$ for $1 \le n \le N$; by the choice of ϵ we have that $d_n(x_n, y_n) < \epsilon_n$ for $1 \le n \le N$. This implies $y \in \bigcap_{k=1}^{N} \pi_{n_k}^{-1}(B(x_{n_k}; \epsilon_k) \subseteq G$. That is, $B(x; \epsilon) \subseteq G$, and hence we have that $G \in \mathcal{D}$.

Now assume $D \in \mathcal{D}$ and $x = \{x_n\} \in D$; pick $\epsilon > 0$ such that $B(x; \epsilon) \subseteq D$; choose $N \ge 1$ such that $\sum_{n=N+1}^{\infty} 2^{-n} < \epsilon/2$. Let $\delta > 0$ such that when $0 \le t < \delta$, $t(1+t)^{-1} < \epsilon/2$. Thus, if $y = \{y_n\} \in X$ and $d_n(x_n, y_n) < \delta$ for $1 \le n \le N$, then

$$\sum_{n=1}^{N} \frac{1}{2^n} \frac{d_n(x_n, y_n)}{1 + d_n(x_n, y_n)} < \sum_{n=1}^{N} \frac{1}{2^n} \frac{\epsilon}{2} < \frac{\epsilon}{2},$$

so that $d(x,y) < \epsilon$. That is, $\bigcap_{n=1}^{N} \pi_n^{-1}(B(x_n; \delta)) \subseteq B(x; \epsilon)$, and we have that $D \in \mathcal{T}$. ∎

The preceding theorem has something to say about Exercise 2.1.10.

Now for one of the deeper results on infinite products. It is one that you will use often if your study of mathematics persists.

Theorem 2.6.7 (Tikhonov's[3] Theorem). *If $\{X_i : i \in I\}$ is a collection of topological spaces and their product X is given the product topology, then X is compact if and only if each X_i is compact.*

Proof. Since each projection map is continuous, it follows that when X is compact, so are all the coordinate spaces. Now assume each X_i is compact and that $\mathcal{C}$ is an open cover of X by elements of the standard subbase $\mathcal{S}$. By Alexander's Theorem, to show that X is compact, it suffices to show that each such cover has a finite subcover.

Let $\mathcal{T}_i$ be the topology on X_i, and for each i let $\mathcal{C}_i = \{G \in \mathcal{T}_i : \pi_i^{-1}(G) \in \mathcal{C}\}$. It is important for later in the proof that we observe that since each

---

[3] Andrei Nikolaevich Tikhonov was born in 1908 in Smolensk, Russia. (His name is often written Tychonoff.) He entered Moscow State University in 1922 and published his first paper in 1925 while still an undergraduate. He received his doctorate in 1927 and was appointed to the faculty of the university in 1933. He first proved the present theorem for the product of an arbitrary infinity of copies of the unit interval and in 1935 stated the full result with the comment that the proof was the same as in the special case. In 1936 he received his habilitation for work on Volterra functional equations and then was made Professor at Moscow State University. Three years later he became a Corresponding Member of the USSR Academy of Sciences. His work now concentrated on differential equations and mathematical physics. In 1966 he was awarded the Lenin Prize and was elected to full membership in the Soviet Academy of Sciences. He had a long and distinguished career including administrative positions as Dean of the Faculty of Computing and Cybernetics at Moscow State University and later as Deputy Director of the Institute of Applied Mathematics of the USSR Academy of Sciences. He died in 1993.

element of the cover $\mathcal{C}$ belongs to the standard subbase $\mathcal{S}$,

2.6.8 $$\mathcal{C} = \bigcup_i \{\pi_i^{-1}(G) : G \in \mathcal{C}_i\}.$$

Claim. There is at least one i in I such that $\mathcal{C}_i$ is a cover of X_i.

In fact, suppose this is not the case. It follows that for every i there is a point x_i that belongs to $X_i \backslash \bigcup \{G : G \in \mathcal{C}_i\}$; let x be the element of X defined by $x(i) = x_i$ for all i. Hence for every i, $x \notin \pi_i^{-1}(G)$ for every G in $\mathcal{C}_i$. But from (2.6.8) it then follows that $x \notin \bigcup \{C : C \in \mathcal{C}\}$. That is, $\mathcal{C}$ is not a cover, a contradiction. Therefore, the claim is established.

Fix an i such that $\mathcal{C}_i$ is a cover of X_i. Since X_i is compact, there are finitely many sets $G_1, \ldots, G_n$ in $\mathcal{C}_i$ such that $X_i = \bigcup_{k=1}^n G_k$. It follows that $X = \bigcup_{k=1}^n \pi_i^{-1}(G_k)$, and so $\{\pi_i^{-1}(G_1), \ldots, \pi_i^{-1}(G_n)\}$ is a finite subcover of $\mathcal{C}$. ∎

Now we want to prove that the product of connected spaces is connected. First we need a lemma.

Lemma 2.6.9. *If X is the product of the topological spaces $\{X_i : i \in I\}$ and $a \in X$, then $D = \{x \in X : x(i) = a(i) \text{ for all but a finite number of } i\}$ is dense in X.*

Proof. If $y \in X$ and G is a neighborhood of y, then the definition of the product topology says that there are $i_1, \ldots, i_n$ in I and open sets G_{i_k} in X_{i_k} such that $y \in \bigcap_{k=1}^n \pi_{i_k}^{-1}(G_{i_k}) \subseteq G$. If we define x in X by setting $x(i) = a(i)$ for $i \neq i_1, \ldots, i_n$ and $x(i_k)$ equal to any point in G_{i_k} for $1 \leq k \leq n$, then $x \in G \cap D$. Therefore, D is dense (Proposition 2.1.9). ∎

Theorem 2.6.10. *If $\{X_i : i \in I\}$ is a collection of topological spaces and their product X is given the product topology, then X is connected if and only if each X_i is connected.*

Proof. If X is connected, then each X_i is connected since it is the image of X under the projection map. Now assume that each X_i is connected. Let $f : X \to \{0, 1\}$ be a continuous function. According to Exercise 2.4.5, it suffices to show that such a continuous function is constant. So fix a point a in X and let us show that $f(x) = f(a)$ for all x in X. If $k \in I$, define $g_k : X_k \to X$ by letting

$$g_k(y)(i) = \begin{cases} y & \text{if } i = k, \\ a(i) & \text{if } i \neq k. \end{cases}$$

We will show that g_k is continuous by showing that $g_k^{-1}(S)$ is open in X_k for every element S of the standard subbasis for the topology on X; that is, we will show that for any i in I $g_k^{-1}(\pi_i^{-1}(U_i))$ is open in X_k whenever U_i is open in X_i. (This will call for some mental dexterity, so pay attention.)

To begin, we show that if U_k is open in X_k, then $g_k^{-1}(\pi_k^{-1}(U_k))$ is open in X_k. Note that $x \in \pi_k^{-1}(U_k)$ if and only if $x(k) \in U_k$. Thus, for y in X_k,

$g_k(y) \in \pi_k^{-1}(U_k)$ if and only if $y = g_k(y)(k) \in U_k$. That is, $g_k^{-1}(\pi_k^{-1}(U_k)) = U_k$, which is open. Now assume $i \neq k$ and U_i is open in X_i. Here $x \in \pi_i^{-1}(U_i)$ if and only if $x(i) \in U_i$. For $y \in X_k$, $g_k(y) \in \pi_i^{-1}(U_i)$ if and only if $a(i) = g_k(y)(i) \in U_i$. Hence we have that when $a(i) \in U_i$, $g_k^{-1}(\pi_i^{-1}(U_i)) = X_k$; and when $a(i) \notin U_i$, $g_k^{-1}(\pi_i^{-1}(U_i)) = \emptyset$. Therefore, g_k is continuous.

This implies that $f \circ g_k : X_k \to \{0, 1\}$ is continuous; since X_k is connected, this function must be constant. But since $g_k(a(k)) = a$, this means that $f(g_k(y)) = f(a)$ for all y in X_k. Equivalently, for any choice of k in I

2.6.11 $f(x) = f(a)$ whenever $x(i) = a(i)$ for $i \neq k$.

It follows that f is constantly equal to $f(a)$ on $D = \{x \in X : x(i) = a(i)$ for all but a finite number of $i\}$. In fact if $i_1, \ldots, i_n \in I$, $x \in X$, and $x(i) = a(i)$ for $i \neq i_1, \ldots, i_n$, then by taking $k = i_1$ in (2.6.11) we have that $f(x) = f(a)$. Since the preceding lemma says D is dense, and since f was assumed continuous, it must be that f is constant on X. Therefore, X is connected.   ∎

The statement for pathwise connected spaces analogous to the preceding proposition is also valid.

Proposition 2.6.12. *If $\{X_i : i \in I\}$ is a collection of topological spaces and their product X is given the product topology, then X is pathwise connected if and only if each X_i is pathwise connected.*

Proof. Assume each X_i is pathwise connected. Let $x = \{x_i\}$ and $y = \{y_i\}$ be two points in X, and for each i let $f_i : [0, 1] \to X_i$ be a path from x_i to y_i. Define $f : [0, 1] \to X$ by $f(t) = \{f_i(t)\}$. Note that if π_i is the projection of X onto X_i, then $p_i \circ f = f_i$. By Proposition 2.6.5, f is continuous, and clearly it is a path from x to y. The converse follows by Proposition 2.5.6(b).   ∎

Exercises

(1) Show that if $X = \prod_i X_i$ and $y = \{y_i\} \in X$, then for each index j the map $\phi : X_j \to X$ defined by

$$\phi(x)_i = \begin{cases} x & \text{when } i = j \\ y_i & \text{when } i \neq j \end{cases}$$

is a homeomorphism of X_j onto $\phi(X_i)$.

(2) Let (X, d) be a metric space, and let $C(X)$ be the set of all continuous functions from X into $\mathbb{R}$. Show that the weak topology defined on X by the functions in $C(X)$ is the given topology on X defined by the metric.

(3) Let $\{(X_i, \mathcal{T}_i) : i \in I\}$ be a collection of topological spaces, and let $\mathcal{B}$ be the topology on $X = \prod_i X_i$ with a subbase $\{\prod_i G_i : G_i \in \mathcal{T}_i\}$. (This is called the *box topology* on X.) (a) Let $I = \mathbb{N}$, and for each n in $\mathbb{N}$ let $X_n = \mathbb{R}$. Give the resulting product space the box topology and define the function $f : \mathbb{R} \to X$ by $f(x) = (x, x, \ldots)$. Show that f is not

continuous, even though $\pi_n \circ f : \mathbb{R} \to \mathbb{R}$ is continuous for each n. [Hint: examine $f^{-1}\left(\prod_{n=1}^{\infty}(-\frac{1}{n}, \frac{1}{n})\right)$.] (b) Again let $I = \mathbb{N}$, and for each n in $\mathbb{N}$ let $X_n = \{0, 1\}$. Show that the box topology on X is the discrete topology. Hence X is not compact, even though each coordinate space is compact.

(4) Find an example of a collection of connected topological spaces such that the product space with the box topology is not connected.

(5) Let $\{(X_i, \mathcal{T}_i) : i \in I\}$ be a collection of topological spaces, and let $X = \prod_i X_i$ have the product topology. Show that X is separable if and only if I is countable and each X_i is separable.

(6) Let $\{(X_i, \mathcal{T}_i) : i \in I\}$ be a collection of topological spaces, and let $X = \prod_i X_i$ have the product topology. If, for each i in I, C_i is a component of X_i, is $C = \prod_{i \in I} C_i$ a component of X?

(7) Let $\{(X_i, \mathcal{T}_i) : i \in I\}$ be a collection of topological spaces, and let $X - \prod_i X_i$ have the product topology. (a) Show that if X is locally connected, then each X_i is locally connected. (Hint: Proposition 2.4.19 may be useful.) (b) Show that the converse is false by giving a counterexample. (See Exercise 2.4.7.) (c) If you assume that each X_i is locally connected, can you give an additional hypothesis that implies that X is locally connected?

(8) Explore the questions raised in the preceding exercise for local pathwise connectedness.

2.7. Nets

Now we generalize the notion of a sequence with a concept that works almost as well for an arbitrary topological space as sequences work for metric spaces. Two references where the reader will find all this material and more are [6] and [4]. Recall the definition of a partially ordered set given in § A.4.

Definition 2.7.1. A *directed set* is a partially ordered set I with the property that if $i, j \in I$, then there is a k in I with $i, j \le k$.

Example 2.7.2. (a) $\mathbb{N}$ is a directed set.
  (b) The set $\mathbb{Z}$ of all integers is a directed set.
  (c) Let E be any set and order 2^E by inclusion. That is, if $A, B \in 2^E$, then $A \le B$ means $A \subseteq B$. It follows that 2^E is a directed set.
  (d) Again, if E is any set, $\mathcal{F}$ is the collection of all finite subsets of E, and $\mathcal{F}$ is ordered by inclusion, then $\mathcal{F}$ is a directed set.
  (e) Let X be a topological space, and for some x_0 in X let $\mathcal{U}$ denote the collection of all neighborhoods of x_0. It is easy to see that $\mathcal{U}$ is a directed set under reverse inclusion; that is, if $U, V \in \mathcal{U}$, declare $U \ge V$ to mean that $U \subseteq V$.

Let us mention something that is immediate from the definition of a directed set and that will be used often: if I is directed and $i_1, \ldots, i_n \in I$, then there is an i in I with $i \ge i_1, \ldots, i_n$. (Verify!)

Definition 2.7.3. A *net* in a set X is a pair (x, I), where I is a directed set and x is a function from I into X.

We will often (usually) write the net as $\{x_i : i \in I\}$ or $\{x_i\}$ if the directed set I is understood; the function notation $\{x(i) : i \in I\}$ will seldom be used to denote a net. Note that every sequence is a net.

Example 2.7.4. (a) Let 2^E be the directed set from Example 2.7.2(c), and
for each $A \in 2^E$ pick a point x_A in A. Then $\{x_A\}$ is a net.
(b) Again let E be a set, and let $\mathcal{F}$ be the collection of all nonempty finite subsets of $\mathcal{E}$ as in Example 2.7.2(d). If $x_F \in F$ for each F in $\mathcal{F}$, then $\{x_F\}$ is a net.
(c) If E is any set, $\mathcal{F}$ is as in Example 2.7.2(d), and $f : E \to \mathbb{R}$ is a function, then $\{\sum_{e \in F} f(e) : F \in \mathcal{F}\}$ is a net. Technically we define $x : \mathcal{F} \to \mathbb{F}$ by $x(F) = \sum_{e \in F} f(e)$, and $(x, \mathcal{F})$ is a net.
(d) If we consider Example 2.7.2(e) and for each U in $\mathcal{U}$ pick a point x_U in U, then $\{x_U : U \in \mathcal{U}\}$ is a net.

Definition 2.7.5. If $\{x_i\}$ is a net in a topological space X, say that the net *converges* to x, in symbols $x_i \to x$ or $x = \lim_i x_i$, if for every open set G containing x there is an i_0 in I such that $x_i \in G$ for all $i \geq i_0$. Say that $\{x_i\}$ *clusters* at x, in symbols $x_i \to_{\mathrm{cl}} x$, if for every open set G containing x and for every j in I there is an $i \geq j$ with x_i in G.

It is easy to see that if we have a sequence, then this definition of convergence is the same as the concept of a convergent sequence (Exercise 1). Similarly, a sequence clusters at a point x in this sense if and only if it clusters in the sense of sequences (Exercise 1).

Example 2.7.6. The net $\{x_U\}$ defined in Example 2.7.4(d) converges to x_0.

The proof of the next proposition is straightforward (Exercise 2) and a good opportunity to fix the ideas in your mind.

Proposition 2.7.7. (a) *If a net in a topological spaces converges to x, then it clusters at x.*
(b) *A net can converge to only one point.*

Here is one way to define a subnet that extends the idea of a subsequence. If I is a directed set, say that a subset J is *cofinal* if for every i in I there is a j in J with $j \geq i$. It is easily shown that when J is a cofinal subset, then it too is a directed set. So if we have a net $x : I \to X$, then we could define a subnet to be the restriction of x to some cofinal subset. However, this is NOT the definition of a subnet. The more complicated notion of a subnet is formulated in such a way that various results about sequences and subsequences in a metric space can be extended to nets and subnets in a topological space. For example: if a net clusters at x, then it has a subnet that converges to x; a topological space X is compact if and only if every net in X has a convergent subnet.

The interested reader can consult [4] and [6] for the accepted definition of a subnet. We will not use subnets in this book. I have found that I can usually avoid them and have decided to live my life that way. I am not saying that all the readers should follow my example, but for this introduction we will follow the path of least resistance.

Proposition 2.7.8. *Let X and Z be topological spaces.*

(a) *If $f : X \to Z$ and $x \in X$, then f is continuous at x if and only if whenever $x_i \to x$ in X, $f(x_i) \to f(x)$ in Z.*

(b) *If $f : X \to Z$ is continuous at x and $x_i \to_{cl} x$, then $f(x_i) \to_{cl} f(x)$.*

(c) *A subset F of X is closed if and only if whenever we have a net $\{x_i\}$ of points in F that converges to a point x, we have that $x \in F$.*

Proof. (a) Suppose f is continuous at x and $x_i \to x$. If W is a neighborhood of $f(x)$ in Z, then there is a neighborhood G of x with $G \subseteq f^{-1}(W)$. Thus, there is an i_0 such that $x_i \in G$ for all $i \geq i_0$; this says that $f(x_i) \in f(G) \subseteq W$ for all $i \geq i_0$. Since W was arbitrary, this implies $f(x_i) \to f(x)$.

Now assume the stated condition is satisfied, and let us show f is continuous at x. We must show that for every neighborhood W of $f(x)$ there is a neighborhood G of x with $G \subseteq f^{-1}(W)$. Suppose this is not the case. Then there is a neighborhood W of $f(x)$ such that if $\mathcal{T}_x$ is the set of neighborhoods of x in X, then $G \backslash f^{-1}(W) \neq \emptyset$ for every G in $\mathcal{T}_x$; for each G in $\mathcal{T}_x$ let $x_G \in G \backslash f^{-1}(W)$. Now $\mathcal{T}_x$ is a directed set under inclusion (Verify!), and we have that $x_G \to x$, as in Example 2.7.6. However, $\{f(x_G) : G \in \mathcal{T}_x\}$ does not converge to $f(x)$ (Why?), giving the desired contradiction.

(b) The proof of this is similar to the first part of the proof of (a) and is Exercise 3.

(c) Assume F is closed, $\{x_i\}$ is a net in F, and $x_i \to x$. If G is a neighborhood of x, then there is an i_0 such that $x_i \in G$ for $i \geq i_0$. This says that $G \cap F \neq \emptyset$ for every neighborhood G of x, hence $x \in \operatorname{cl} F = F$. Now assume the stated condition holds and $x \in \operatorname{cl} F$. Let $\mathcal{T}_x$ be the set of all neighborhoods of x and order it by inclusion. Since $x \in \operatorname{cl} F$, for every G in $\mathcal{T}_x$ there is a point x_G in $G \cap F$. But $\{x_G : G \in \mathcal{T}_x\}$ is a net that converges to x, so $x \in F$ by assumption. Thus $\operatorname{cl} F \subseteq F$, and F is closed. ∎

Also see Exercise 4 for an extension of part (c) of the preceding proposition. The plot thickens.

Theorem 2.7.9. *A topological space X is compact if and only if every net in X has a cluster point.*

Proof. Suppose X is compact and $\{x_i\}$ is a net in X. For every j in I let F_j be the closure of $\{x_i : i \geq j\}$. If $j_1, \ldots, j_n \in I$, then there is an $i \geq j_k$

for $1 \leq k \leq n$; hence $x_i \in \bigcap_{k=1}^{n} F_{j_k}$. So $\{F_i : i \in I\}$ has the FIP; since X is compact, there is an x in $\bigcap_i F_i$. If G is an open set containing x and $i_0 \in I$, then $x \in F_{i_0}$, so that $G \cap \{x_i : i \geq i_0\} \neq \emptyset$. That is, there is an $i \geq i_0$ with $x_i \in G$. By definition, this says that $x_i \to_{\mathrm{cl}} x$.

Now assume X is not compact, and let $\mathcal{G}$ be an open cover of X without a finite subcover. Let I be the set of all finite subsets of $\mathcal{G}$ and order I by inclusion as in Example 2.7.2(d). By assumption, for every $i = \{G_1, \ldots, G_n\}$ in I there is a point x_i in X such that $x_i \notin \bigcup_{k=1}^{n} G_k$; $\{x_i\}$ is a net in X. This net does not have a cluster point. In fact, if there is a cluster point x, then there is a G in $\mathcal{G}$ that contains x. But $\{G\} \in I$, and so there is an $i = \{G, G_1, \ldots, G_n\} \geq \{G\}$ with $x_i \in G$. But by definition, $x_i \notin G \cup \bigcup_{k=1}^{n} G_k$, a contradiction. ∎

The next theorem is very useful.

Theorem 2.7.10. *If X is a compact space and $\{x_i : i \in I\}$ is a net in X with a unique cluster point x, then $\{x_i\}$ converges to x.*

Proof. Fix a proper open set G containing x. If it were the case that the net did not converge to x, then for every i_0 in I, there would be an i in I with $i \geq i_0$ and x_i in $X \backslash G$. This says that $J = \{j \in I : x_j \in X \backslash G\} \neq \emptyset$. In fact, this also says that J is directed. Indeed, if $j_1, j_2 \in J$, let $i_0 \in I$ such that $i_0 \geq j_1, j_2$. By what we have said, there is a j in J with $j \geq i_0 \geq j_1, j_2$. Thus, $\{x_j : j \in J\}$ is a net in $X \backslash G$. Since $X \backslash G$ is compact, there is a y in $X \backslash G$ such that $\{x_j : j \in J\} \to_{\mathrm{cl}} y$. The reader can check that this implies that y is also a cluster point of the original net. Since $y \neq x$, we have a contradiction. ∎

Exercises

(1) Show that if $\{x_n\}$ is a sequence in a topological space, then $x_n \to x$ as a sequence if and only if $x_n \to x$ as a net. Show that $x_n \to_{\mathrm{cl}} x$ if and only if x is a limit point of $\{x_1, x_2, \ldots\}$.

(2) Prove Proposition 2.7.7.

(3) Prove Proposition 2.7.8(b). Is the converse true?

(4) Show that a subset F of X is closed if and only if whenever $\{x_i\}$ is a net in F and x is a cluster point of this net, it follows that $x \in F$.

(5) If (X, d) is a metric space, $\{x_i\}$ is a net in X, and x is a cluster point of this net, show that there are $\{i_n : n \geq 1\}$ such that $i_1 \leq i_2 \leq \cdots$ and $x = \lim_n x_{i_n}$.

(6) Let $\mathcal{S}$ be a subbase for the topology on X. (a) Show that a net $\{x_i\}$ in X converges to x if and only if for every S in $\mathcal{S}$ that contains x, there is an i_0 such that $x_i \in S$ for all $i \geq i_0$. (b) Find an example of a topological space X, a subbase $\mathcal{S}$ for the topology of X, a net $\{x_i\}$ in X, and a point x such that for each S in $\mathcal{S}$ that contains x and each i there is a $j \geq i$ such that $x_j \in S$, but the net $\{x_i\}$ does not cluster at x.

(7) Let $\{X^\alpha : \alpha \in A\}$ be a family of topological spaces, let $X = \prod_\alpha X^\alpha$, let $\{x_i : i \in I\}$ be a net in X with each $x_i = \{x_i^\alpha\}$, and let $x = \{x^\alpha\} \in X$. (a) Show that $x_i \to x$ if and only if $x_i^\alpha \to x^\alpha$ for each α in A. (b) Find an example of a sequence $\{x_n\}$ in $\mathbb{R}^2$ with $x_n = (x_n^1, x_n^2)$ and a point $x = (x^1, x^2)$ such that $x_n^1 \to_{\text{cl}} x^1$ and $x_n^2 \to_{\text{cl}} x^2$ but $\{x_n\}$ does not cluster at x.

(8) Find an example of a net in a topological space X that has a unique cluster point but that does not converge. See Theorem 2.7.10.

2.8. Quotient Spaces

We begin this section by examining the concept of an equivalence relation on a set, which is the starting point for discussing quotient spaces. I suspect some readers, perhaps many, have already encountered quotient spaces, though maybe not in their abstract formulation. Quotient spaces are frequently covered in a first course in linear algebra, for example. Nevertheless, this concept often causes students difficulty, so let us take the time to present it carefully.

Definition 2.8.1. An *equivalence relation* on a set X is a relation $\sim$ between elements of X that satisfies the following properties: (reflexivity) $x \sim x$ for all x in X; (symmetry) if $x \sim y$, then $y \sim x$; (transitivity) if $x \sim y$ and $y \sim z$, then $x \sim z$.

By the way, if you want a formal definition of a relation, it is just a subset E of the cartesian product $X \times X$, and we write $x \sim y$ when $(x, y) \in E$. When we have an equivalence relation, this imposes additional restrictions on the set E. As we progress through examples, you and a classmate, perhaps over a cup of coffee or a nice glass of Merlot, might ask what is a necessary and sufficient condition on a subset E of the product space for the corresponding relation to be an equivalence relation. In this book, we will use the definition of an equivalence relation given previously and avoid discussing the subset E of $X \times X$.

Example 2.8.2. (a) Equality is an equivalence relation.
  (b) If $\mathcal{X}$ is a vector space over $\mathbb{R}$, $\mathcal{M}$ is a vector subspace, and we define $x \sim y$ to mean that $x - y \in \mathcal{M}$, then $\sim$ is an equivalence relation.
  (c) (Assuming you know the definition of a group) If G is a group, H is a subgroup, and we define $x \sim y$ to mean that $xy^{-1} \in H$, then this is an equivalence relation.
  (d) If X and Z are two sets, $f : X \to Z$, and we define $x \sim y$ to mean that $f(x) = f(y)$, then this is an equivalence relation.
  (e) Let X be a set, and let $A \subseteq X$. Define $\sim$ as follows: (i) if $x, y \in A$, $x \sim y$; (ii) if $x \notin A$, then the only point y such that $y \sim x$ is $y = x$.

A *partition* of a set X is a collection $\mathcal{P}$ of nonempty subsets of X such that $P \cap R = \emptyset$ for distinct elements P and R of $\mathcal{P}$ and $\bigcup\{P : P \in \mathcal{P}\} = X$. The proof of the next proposition is Exercise 3.

Proposition 2.8.3. *If X is a set with an equivalence relation and for each x in X we let $P_x = \{y \in X : x \sim y\}$, then the collection $\{P_x : x \in X\}$ is a partition of X. Conversely, if $\mathcal{P}$ is a partition of X and we define $x \sim y$ to mean that there is a P in $\mathcal{P}$ such that $x, y \in P$, then $\sim$ is an equivalence relation on X.*

So this gives us another method of defining an equivalence relation—just specify a partition. We will often talk of an equivalence relation and its associated or corresponding partition, or we will talk about a partition and its corresponding equivalence relation. For an equivalence relation on a set, the sets in the corresponding partition are called the *equivalence classes* of the relation.

Definition 2.8.4. When $\sim$ is an equivalence relation on X, we define the quotient space $X/\sim$ to be the collection of equivalence classes. If the equivalence relation is defined by a partition $\mathcal{P}$, we might denote the quotient space by $X/\mathcal{P}$. The map $q : X \to X/\sim$ defined by letting $q(x)$ be the equivalence class that contains x is called the *quotient map* or the *natural map*.

Note that if $\xi \in X/\sim$, then $q^{-1}(\xi)$ is the set $\{x \in X : x \in \xi\}$. Actually, this is just the set ξ, but viewed as a subset of X rather that an element of the quotient space. So if we consider the equivalence relation defined in Example 2.8.2(b), when $x \in \mathcal{X}$, we have that $q(x) = x + \mathcal{M}$.

Now let us look at what happens when we have an equivalence relation on a topological space. Unfortunately, it is not a pretty picture. We will define a topology on the quotient space, except that it will not always satisfy the Hausdorff property. There are some abstract conditions under which this topology is Hausdorff, though I have never had to resort to them to prove a quotient space Hausdorff—the quotient spaces I have encountered are usually clearly Hausdorff or clearly they are not. There are two points worth making about this. First, many quotient spaces are Hausdorff, and it is not so difficult to verify this. (See Example 3.1.7 in the next chapter. Also, Exercise 3.1.7 gives an example of a quotient space that is not Hausdorff.) Second, even when the quotient space is Hausdorff, it may not inherit other topological properties possessed by X.

The proof of the following basic result is straightforward (Exercise 5).

Proposition 2.8.5. *Let X be a topological space with an equivalence relation $\sim$. If $q : X \to X/\sim$ is the quotient map, then $\mathcal{U} = \{U \subseteq X/\sim : q^{-1}(U)$ is open in $X\}$ is a possibly non-Hausdorff topology on $X/\sim$ and q becomes a continuous mapping.*

The topology $\mathcal{U}$ defined in the preceding proposition is called the *quotient topology* on $X/\sim$. Whenever X is a topological space and we have an equivalence relation on X, it will always be assumed that when we discuss topological ideas on the quotient space we are discussing the quotient topology. Note that because the quotient map is continuous, it follows that when X is either compact or connected, so is $X/\sim$.

Example 2.8.6. (a) Let X be the closed unit square in the plane: $X = [0, 1] \times [0, 1]$. Define an equivalence relation $\sim$ on X by identifying the two vertical sides. Precisely, $(0, y) \sim (1, y)$ for $0 \leq y \leq 1$, while the remaining points are identified only with themselves. The quotient space $X/\sim$ is homeomorphic to a closed hollow cylinder in $\mathbb{R}^3$.

(b) Let X be as in part (a), but define the equivalence relation $\sim$ by identifying the two vertical sides and by identifying the two horizontal sides. Precisely, (i) define $(0, y) \sim (1, y)$ for $0 \leq y \leq 1$; (ii) define $(x, 0) \sim (x, 1)$ for $0 \leq x \leq 1$; (iii) apply the transitive law to these relations to get $(x, y) \sim (w, z)$ when $x, y, w, z \in \{0, 1\}$; (iv) the remaining points are identified only with themselves. Show that the quotient space $X/\sim$ is homeomorphic to the hollow torus (or doughnut) in $\mathbb{R}^3$.

Proposition 2.8.7. *If X and Z are topological spaces, $\sim$ is an equivalence relation on X with quotient map $q : X \to X/\sim$, and $f : X/\sim \to Z$, then f is continuous if and only if $f \circ q : X \to Z$ is continuous.*

Proof. If f is continuous, then $f \circ q$ is the composition of two continuous maps and so must be continuous. If $f \circ q$ is continuous and V is an open subset of Z, then $q^{-1}[f^{-1}(V)] = (f \circ q)^{-1}(V)$ and is therefore open in X. By definition, this implies $f^{-1}(V)$ is open in $X/\sim$. Therefore, f is continuous. ∎

As we proceed through the rest of the book and encounter new properties of a topological space, we will address the issue of whether these properties are preserved by taking quotients.

Exercises

(1) Verify that each of the relations in Example 2.8.2 is an equivalence relation.

(2) For each of the equivalence relations in Example 2.8.2 describe the corresponding partition of the underlying set X as in Proposition 2.8.3.

(3) Prove Proposition 2.8.3.

(4) For each equivalence relation in Example 2.8.2, describe the corresponding partition, the quotient space, and the natural map.

(5) Prove Proposition 2.8.5.

(6) Let X be a topological space with an equivalence relation $\sim$ such that $X/\sim$ is a Hausdorff space. If X is locally connected, show that $X/\sim$ is locally connected.

(7) Define a *semimetric* on a set X to be a function $\rho : X \times X \to [0, \infty)$ satisfying the following conditions for all x, y, z in X: (i) $\rho(x, x) = 0$; (ii) $\rho(x, y) = \rho(y, x)$; (iii) $\rho(x, y) \leq \rho(x, z) + \rho(z, y)$. If ρ is a given semimetric on X, define an equivalence relation on X by $x \sim y$ when $\rho(x, y) = 0$. (a) Verify that this is an equivalence relation, and describe the equivalence classes. (b) If $q : X \to X/\sim$ is the natural map, show

that $d(q(x), q(y)) = \rho(x, y)$ is a well-defined metric on $X/\sim$. (c) Show that the metric defined in (b) defines the quotient topology on $X/\sim$.

(8) If X is a pathwise connected space and $\sim$ is an equivalence relation on X such that $X/\sim$ is Hausdorff, show that $X/\sim$ is pathwise connected.

(9) (a) Prove the statement in Example 2.8.6(a). (b) Prove the statement in Example 2.8.6(b). (c) What happens in Example 2.8.6 if all the points on the boundary of X are identified to a single point?

Chapter 3

Continuous Real-Valued Functions

In this chapter we will focus on continuous functions from a topological space into the space of real numbers, $\mathbb{R}$. We have already seen some such results in §2.3, but in this chapter we will extend the exposure and deepen the probing.

3.1. Convergence of Functions

We want to start thinking of functions as points in a topological space. For any topological space X let $C(X)$ denote the vector space of all continuous functions from X into $\mathbb{R}$. [Recall Proposition 2.3.3, where it is shown that $C(X)$ is closed under sums and products. So $C(X)$ is not only a vector space but also an algebra. We also refer the reader to Exercise 1.3.13.] A function $f : X \to \mathbb{R}$ is said to be *bounded* if there is a constant M with $|f(x)| \leq M$ for all x in X. Let $C_b(X)$ denote the space of all bounded continuous functions from X into $\mathbb{R}$. It is easy to check that $C_b(X)$ is also closed under forming sums and products so that it too is an algebra. Of course, when X is compact, $C(X) = C_b(X)$.

The constant functions belong to $C(X)$, but are there any nonconstant functions? That is a problem, and in subsequent sections we will address this question and obtain satisfactory answers. For now suffice it to say that when X is a metric space, we know from Urysohn's Lemma (Theorem 1.3.9) that $C_b(X)$ has a rich supply of continuous functions.

If $f, g \in C_b(X)$, define

3.1.1 $$\rho(f, g) = \sup\{|f(x) - g(x)| : x \in X\}.$$

J.B. Conway, *A Course in Point Set Topology*, Undergraduate Texts in Mathematics, DOI 10.1007/978-3-319-02368-7_3,
© Springer International Publishing Switzerland 2014

Proposition 3.1.2. *The function defined on $C_b(X) \times C_b(X)$ in (3.1.1) is a metric on $C_b(X)$.*

Proof. If $f, g, h \in C_b(X)$ and $x \in X$, then $|f(x) - g(x)| \leq |f(x) - h(x)| + |h(x) - g(x)|$. Hence $\rho(f, g) \leq \sup\{|f(x) - h(x)| + |h(x) - g(x)| : x \in X\} \leq \sup\{|f(x) - h(x)|; x \in X\} + \sup\{|h(x) - g(x)| : x \in X\} = \rho(f, h) + \rho(h, g)$. This establishes the triangle inequality. The proof that the other properties of a metric hold is routine. ∎

Whenever we discuss the topological properties of $C_b(X)$, it is assumed we are discussing the topology defined by this metric. The preceding proposition is the reason we made the statement earlier that the reader should start thinking of functions in $C_b(X)$ as points; they are indeed points in a metric space. Also, see Exercise 8, where a topology is defined on $C(X)$ when X is not assumed to be compact.

Definition 3.1.3. If $\{f_n\}$ and f are bounded functions from a set X into $\mathbb{R}$, then say that $\{f_n\}$ *converges uniformly* to f if for every $\epsilon > 0$ there is an N such that $|f_n(x) - f(x)| < \epsilon$ for all x in X and all $n \geq N$. Say that $\{f_n\}$ is a *uniformly Cauchy sequence* if for every $\epsilon > 0$ there is an integer N such that $|f_n(x) - f_m(x)| < \epsilon$ for all x in X and all $m, n \geq N$.

This notion is possibly introduced in basic courses on calculus when the functions are defined on a subset of the real line. In any case, we have the following proposition.

Proposition 3.1.4. *Let X be a topological space, and let $\{f_n\}$ be a sequence in $C_b(X)$. If $f \in C_b(X)$, then $\{f_n\}$ converges to f in the metric of $C_b(X)$ if and only if the sequence converges uniformly to f. The sequence $\{f_n\}$ is a Cauchy sequence in the metric space $C_b(X)$ if and only if it is a uniformly Cauchy sequence.*

Proof. The proof is straightforward. If $\rho(f_n, f) \to 0$ and $\epsilon > 0$, then there is an N with $\rho(f_n, f) < \epsilon$ for $n \geq N$. By the definition of the metric, this implies $|f_n(x) - f(x)| < \epsilon$ for $n \geq N$ for all x in X; that is, $f_n \to f$ uniformly on X. Now assume we have uniform convergence, $\epsilon > 0$ is given, and N is as in the definition. Formula 3.1.1 implies $\rho(f_n, f) \leq \epsilon$ for $n \geq N$, and so $f_n \to f$ in $C_b(X)$.

The proof regarding Cauchy sequences is similar to the last paragraph and is left to the reader. ∎

Theorem 3.1.5. *For any topological space X, $C_b(X)$ is a complete metric space.*

Proof. Let $\{f_n\}$ be a Cauchy sequence in $C_b(X)$. The definition of the metric implies that for each x in X, $|f_n(x) - f_m(x)| \leq \rho(f_n, f_m)$. Therefore, for each x in X, $\{f_n(x)\}$ is a Cauchy sequence in $\mathbb{R}$; define $f(x) = \lim_n f_n(x)$. The task now is to show that $f \in C_b(X)$ and $\rho(f_n, f) \to 0$. Pay heed to the argument used here because similar arguments occur frequently.

The fact that $f : X \to \mathbb{R}$ is a bounded function is the easiest part. Let $n \geq 1$ such that $\rho(f_n, f_m) < 1$ for $m, n \geq N$. Since each of the functions $f_1, \ldots, f_N$ is bounded , there is one constant M such that $|f_k(x)| \leq M$ for $1 \leq k \leq N$ and all x in X. If $n > N$ and $x \in X$, then $|f_n(x)| \leq \rho(f_n, f_N) + |f_N(x)| < 1 + M$. Hence for each x the sequence $\{f_n(x)\} \subseteq [-(M+1), M+1]$, and so $|f(x)| \leq M + 1$; that is, f is a bounded function.

Once again, fix an $\epsilon > 0$, and now let N be such that $\rho(f_n, f_m) < \epsilon/3$ for all $m, n \geq N$. If x is an arbitrary point in X and $n \geq N$, then $|f(x) - f_n(x)| \leq |f(x) - f_m(x)| + \rho(f_m, f_n) < |f(x) - f_m(x)| + \epsilon/3$ for all $m \geq N$. Letting $m \to \infty$, we get that $|f(x) - f_n(x)| \leq \epsilon/3$ for all $n \geq N$ and an arbitrary x. Thus, $\sup\{|f(x) - f_n(x)| : x \in X\} \leq \epsilon/3$ for all $n \geq N$. Note that once we establish that f is continuous, this shows that $\rho(f_n, f) \to 0$.

To show that f is a continuous function, we continue the notation from the preceding paragraph. Fix x and fix $n \geq N$. Since f_n is continuous, there is a neighborhood G of x such that $|f_n(y) - f_n(x)| < \epsilon/3$ for all y in G. Hence, when $y \in G$, $|f(y) - f(x)| \leq |f(y) - f_n(y)| + |f_n(y) - f_n(x)| + |f_n(x) - f(x)| < \epsilon$. Therefore, $f \in C_b(X)$; as we mentioned before, this shows that $f_n \to f$ in the metric space $C_b(X)$, and $C_b(X)$ is complete. ∎

Recall the ordering of the functions in $C_b(X)$. I have always found the next result intriguing; on the other hand, I have never used it. See Exercise 6.

Proposition 3.1.6 (Dini's[1] Theorem). *If X is compact, $\{f_n\}$ is an increasing sequence in $C(X)$, and $f \in C(X)$ such that $f_n(x) \to f(x)$ for all x in X, then $f_n \to f$ uniformly on X.*

Proof. Let $\epsilon > 0$, and for each $n \geq 1$ let $U_n = \{x \in X : f(x) < f_n(x) + \epsilon\}$. Because the functions f_n are increasing with n, it follows that $U_n \subseteq U_{n+1}$ for all $n \geq 1$. Since $f_n(x) \to f(x)$ for every x in X, we have that $X = \bigcup_{n=1}^{\infty} U_n$. Finally, since both f and f_n are continuous, each U_n is open. Because X is compact, the open cover $\{U_n\}$ has a finite subcover; by the increasing property of these sets, this means there is a single integer N such

---

[1]Ulisse Dini was born in 1845 in Pisa, Italy, where he attended the Scuola Normale Superiore, a teaching preparatory college. In 1865 he won a scholarship for study abroad, which he used to go to Paris for a year. During this time he was very active in research, eventually publishing seven papers based on the work he had done. He returned to Pisa and an academic position at the university. Dini's life span was a period of myriad political developments in Italy as the country worked its way toward unification. This is not a period for the casual historian. In 1859 there was a war with Austria, and in 1861 the Kingdom of Italy was formed, though it did not include Venice and Rome. It was not until 1866 that Venice became part of the kingdom, and Rome had to wait until 1870. The turmoil affected Dini, and he progressed in both his academic and political careers. In 1871 he took over Betti's chair of analysis, and that same year he was elected to the Pisa city council. In 1877 he was appointed to a second chair in mathematics, and in 1880 he was elected as a representative of Pisa to the national assembly. In 1883 he was appointed Rector of the university, holding the position for 2 years. In 1892 he was elected Senator in the Italian parliament, and in 1908 he became Director of the Scuola Normale Superiore, a position he held for the rest of his life. This was a period of development in mathematical analysis when the turmoil seemed to be trying to parody the events in Italy; mathematicians sought rigorous proofs of results that had only casually been established, and they sought the boundaries of validity for these results. Dini seemed to flourish in this undertaking; in addition to the present result, there is one in Fourier series that bears his name. He also wrote several influential texts. He died in 1918 in Pisa.

that $U_N = X$. Thus, for $n \geq N$, $U_n \subseteq U_N$, and we have $f_n(x) \leq f(x) <$ $f_n(x) + \epsilon$. By definition, $f_n \to f$ uniformly. $\blacksquare$

Let us look at an example that illustrates an instance where we have a Hausdorff quotient space.

Example 3.1.7. Let X be a topological space, and assume F is a closed subset. Consider $C_b(X)$, and let $\mathcal{M} = \{f \in C_b(X) : f(x) = 0 \text{ for all } x \in F\}$. Note that $\mathcal{M}$ is a linear subspace of the vector space $C_b(X)$ and that it is a closed subset of the metric space $C_b(X)$. (Verify!) We define an equivalence relation on $C_b(X)$ by $f \sim g$ when $f - g \in \mathcal{M}$ and denote the quotient space by $C_b(X)/\mathcal{M}$. Because of the vector space structure, this is precisely the vector space quotient from linear algebra. So $C_b(X)/\mathcal{M} = \{f + \mathcal{M} : f \in C_b(X)\}$ and $f + \mathcal{M} = g + \mathcal{M}$ if and only if $f - g \in \mathcal{M}$; equivalently, if and only if $f(x) = g(x)$ for all x in F. We want to show that $C_b(X)/\mathcal{M}$ is Hausdorff. To do this, we first show that $q : C_b(X) \to C_b(X)/\mathcal{M}$ is an open mapping.

If Ω is any subset in $C_b(X)$, then we claim that

$$q^{-1}[q(\Omega)] = \Omega + \mathcal{M} = \{h + f : h \in \Omega \text{ and } f \in \mathcal{M}\}.$$

In fact, if $g \in q^{-1}[q(\Omega)]$, then $q(g) \in q(\Omega)$, which means that there is an h in Ω such that $g - h \in \mathcal{M}$; that is, $g \in h + \mathcal{M} \subseteq \Omega + \mathcal{M}$. Conversely, if $g = h + f$ with h in Ω and f in $\mathcal{M}$, then $q(g) = q(h)$, so that $g \in q^{-1}[q(\Omega)]$, and we have that $q^{-1}[q(\Omega)] = \Omega + \mathcal{M}$.

Therefore, if Ω is an open subset of $C_b(X)$, then $q^{-1}[q(\Omega)] = \bigcup\{\Omega + f : f \in \mathcal{M}\}$; since this is the union of open sets, $q(\Omega)$ is open.

Now suppose $f + \mathcal{M} \neq g + \mathcal{M}$. Thus, there is a point x in F such that $f(x) \neq g(x)$; without loss of generality we may assume that $f(x) < g(x)$. Pick any real number a satisfying $f(x) < a < g(x)$ and define $\Omega = \{u \in C_b(X) : u(x) < a\}, \Lambda = \{v \in C_b(X) : v(x) > a\}$. It is left to the reader to show that Ω and Λ are open subsets of $C_b(X)$ with f in Ω and g in Λ. So $q(\Omega)$ and $q(\Lambda)$ are open subsets of $C_b(X)/\mathcal{M}$. If $h + \mathcal{M} \in q(\Omega) \cap q(\Lambda)$, then there must be functions u in Ω and v in Λ such that $h - u, h - v \in \mathcal{M}$. Thus, $h(x) = u(x) < a$ and $h(x) = v(x) > a$, a contradiction. Hence $q(\Omega)$ and $q(\Lambda)$ are disjoint open sets that separate $f + \mathcal{M}$ and $g + \mathcal{M}$. Therefore, the quotient space is Hausdorff.

Exercises

(1) Show that $C_b(\mathbb{N})$ can be identified with the space ℓ^∞, which was examined in Exercises 1.1.12, 1.2.11, and 1.4.11.

(2) Say that a series $\sum_{n=1}^{\infty} f_n$ of functions f_n in $C_b(X)$ converges to f if the sequence of finite sums $\{\sum_{k=1}^{n} f_k\}_n$ converges in the metric of $C_b(X)$ to f. Prove the Weierstrass M-test: if $\{f_n\}$ is a sequence in $C_b(X)$ such that there are constants $\{M_n\}$ with $|f_n(x)| \leq M_n$ for all $n \geq 1$ and $\sum_{n=1}^{\infty} M_n < \infty$, then there is a function f in $C_b(X)$ such that the infinite series $\sum_{n=1}^{\infty} f_n$ converges to f.

(3) If (X, d) is a noncompact metric space, show that there is an unbounded continuous function from X into $\mathbb{R}$.

(4) If (X, d) is a metric space and $C_u(X)$ denotes the set of all bounded uniformly continuous functions from X into $\mathbb{R}$, show that $C_u(X)$ is a closed subset of $C_b(X)$.

(5) Consider $C_b(X)$ and define the usual order on it: $f \leq g$ means $f(x) \leq g(x)$ for all x in X. Give $(C_b(X), \leq)$ the order topology (Exercise 2.2.5). (a) Show that the order topology on $C_b(X)$ is a Hausdorff topology. (b) How does the order topology compare to the topology defined by the metric?

(6) Give an example of a sequence $\{f_n\}$ in $C([0, 1])$ such that $f_n(x) \to 0$ for every x in $[0, 1]$, but where $\{f_n\}$ does not converge uniformly on $[0, 1]$. Note that this does not contradict Dini's Theorem.

(7) Here is an abstraction of part of Example 3.1.7. Let $\mathcal{X}$ be a vector space over $\mathbb{R}$ that has a topology such that: (i) the map $(x, y) \mapsto x + y$ is continuous from $\mathcal{X} \times \mathcal{X} \to \mathcal{X}$; (ii) the map $(a, x) \mapsto ax$ is continuous from $\mathbb{R} \times \mathcal{X} \to \mathcal{X}$. (Such a space $\mathcal{X}$ is called a *topological vector space.*) (a) If $y \in \mathcal{X}$, show that the map of $\mathcal{X}$ into itself defined by $x \mapsto x + y$ is a homeomorphism. Let $\mathcal{M}$ be a vector subspace of $\mathcal{X}$, and consider the quotient subspace $\mathcal{X}/\mathcal{M}$ with quotient map $q : \mathcal{X} \to \mathcal{X}/\mathcal{M}$. (b) Show that the quotient map is an open mapping. (c) Show that $\mathcal{X}/\mathcal{M}$ is Hausdorff if and only if, when $x \notin \mathcal{M}$, there are disjoint open sets U and V in $\mathcal{X}/\mathcal{M}$ such that $\mathcal{M} = q(0) \in U$ and $x + \mathcal{M} = q(x) \in V$. (d) Show that if $\mathcal{M}$ is not a closed subset of $\mathcal{X}$, then $\mathcal{X}/\mathcal{M}$ cannot be Hausdorff. (e) If we take $\mathcal{X} = C_b(\mathbb{N})$ and $\mathcal{M} = \{f \in C_b(\mathbb{N}) : \text{there is an integer } N \geq 2 \text{ with } f(n) = 0 \text{ for } n \leq N\}$, show that $\mathcal{M}$ is a linear subspace of $\mathcal{X}$. Is $\mathcal{X}/\mathcal{M}$ Hausdorff?

(8) Let X and Z be topological spaces, and denote by $C(X, Z)$ the set of all continuous functions from X into Z. When K is a compact subset of X and G is an open subset of Z, let

$$\Omega(K, G) = \{h \in C(X, Z) : h(K) \subseteq G\}.$$

(a) Show that the collection of all such sets $\Omega(K, G)$ is a subbase for a topology on $C(X, Z)$. This topology is called the *compact-open topology,* denoted by (co). (b) If the space X consists of a single point x_0, show that there is a homeomorphism between Z and the space $C(\{x_0\}, Z)$ with (co). (c) Show that $C([0, 1], [0, 1])$ with (co) is not a compact space. (Hint: examine the sequence $\{t^n\}$ in $C([0, 1], [0, 1])$.) (d) Let $Y = \mathbb{R}$ so that $C(X, Y) = C(X)$. For each f in $C(X)$, each compact subset K of X, and each $\epsilon > 0$, let

$$S_{K, \epsilon}(f) = \{g \in C(X) : \sup\{(f(x) - g(x)| : x \in K\} < \epsilon\}.$$

Show that the collection of all the sets $S_{K, \epsilon}(f)$ is a subbase for the (co) on $C(X)$. (e) If X is a compact space, show that the (co) topology $C(X)$ is the same as the topology defined by the metric ρ given at the start of this section. (f) Show that a net $\{f_i\}$ in $C(X)$ converges in (co) to f if and only if for every compact subset K of X, $f_i(x) \to f(x)$ uniformly on K.

3.2. Separation Properties

Here we begin to explore various ways of separating disjoint closed subsets of a topological space by disjoint open sets as well as by functions in $C(X)$. In a sense, we incorporated one such separation into the definition of a topological space when we assumed all spaces had the Hausdorff property, which says that singleton sets can be separated. In this section, we introduce two levels of separation, and in the next section we will take separation to a further level.

Definition 3.2.1. A topological space X is *regular* if for every closed set F and any point x not belonging to F there are disjoint open sets U and V such that $x \in U$ and $F \subseteq V$.

We are not that interested in regular spaces, though we will prove some results about them below. Suffice it to say that there are topological spaces that are not regular (Exercise 1), but they tend to be pathological.

Proposition 3.2.2. *If X is a topological space, then the following statements are equivalent.*

(a) X is regular.

(b) If G is an open set and $x \in G$, then there is an open set U such that $x \in U \subseteq \operatorname{cl} U \subseteq G$.

(c) If F is a closed set and $x \notin F$, then there is an open set V such that $F \subseteq V$ and $x \notin \operatorname{cl} V$.

Proof. Assume (a). If G and x are as in the statement of (b), then $F = X \backslash G$ is closed and $x \notin F$. If U and V are disjoint open sets such that $x \in U$ and $F \subseteq V$, then $\operatorname{cl} U \subseteq X \backslash V \subseteq X \backslash F = G$. Now assume (b) holds and x and F are as in (c). Apply (b) with $G = X \backslash F$ to obtain an open set U such that $x \in U \subseteq \operatorname{cl} U \subseteq X \backslash F$. Put $V = X \backslash \operatorname{cl} U$. It follows that $V \subseteq X \backslash U$, so that $\operatorname{cl} V \subseteq X \backslash U$; in particular, $x \notin \operatorname{cl} V$. Now assume (c). If F is closed and $x \notin F$, then (c) implies the existence of an open set V as stated there. Thus, $U = X \backslash \operatorname{cl} V$ is open, $x \in U$, and $U \cap V = \emptyset$. Thus, X is regular. ∎

Proposition 3.2.3. (a)   If X is regular and $E \subseteq X$, then E with its relative topology is regular.

(b)   If $X = \prod_i X_i$, then X is regular if and only if and each X_i is regular.

Proof.   (a) The proof of this part is Exercise 2.

(b) Assume each X_i is regular. If G is an open set in X and $x \in G$, then the definition of the topology on X there are indices $i_1, \ldots, i_n$ and open sets V_{i_k} in X_{i_k} such that $x \in \bigcap_{k=1}^n p_{i_k}^{-1}(V_{i_k}) \subseteq G$. Since each X_{i_k} is regular, Proposition 3.2.2 implies there is an open subset U_{i_k} in X_{i_k} with

$p_{i_k}(x) \in U_{i_k} \subseteq \mathrm{cl}\, U_{i_k} \subseteq V_{i_k}$. Thus, $x \in U = \bigcap_{k=1}^n p_{i_k}^{-1}(U_{i_k}) \subseteq \mathrm{cl}\, U \subseteq$ $\bigcap_{k=1}^n p_{i_k}^{-1}(\mathrm{cl}\, U_{i_k}) \subseteq \bigcap_{k=1}^n p_{i_k}^{-1}(V_{i_k}) \subseteq G$. By the preceding proposition X is regular.

The proof of the converse is Exercise 7.   ∎

Proposition 3.2.4. *Every compact space and every metric space is regular.*

Proof. Assume X is compact, F is a closed subset of X, and $x \in X\backslash F$. For each y in F let U_y and V_y be disjoint open sets in X such that $x \in U_y$ and $y \in V_y$. It follows that $\{V_y : y \in F\}$ is an open cover of F. Since F is compact, there are $y_1, \ldots, y_n$ in Y such that $F \subseteq \bigcup_{k=1}^n V_{y_k}$. Put $V = \bigcup_{k=1}^n V_{y_k}$ and $U = \bigcap_{k=1}^n U_{y_k}$. It follows that U and V are disjoint open sets with x in U and F contained in V.

Now assume we have a metric space (X, d). If F is closed and $x \notin F$, then F and $\{x\}$ are disjoint closed sets. By Urysohn's Lemma (Theorem 1.3.9) there is a continuous function $f : X \to [0, 1]$ with $f(x) = 1$ and $f(y) = 0$ for every y in F. Hence $U = \{y : f(y) > \frac{1}{2}\}$ and $V = \{y : f(y) < \frac{1}{2}\}$ are disjoint open sets that separate x from F.   ∎

The proof just given, that metric spaces are regular, anticipates the next level of separation.

Definition 3.2.5. A topological space X is *completely regular* if for any closed subset F and any point x in $X\backslash F$ there is a continuous function $f : X \to \mathbb{R}$ such that $f(x) = 1$ and $f(y) = 0$ for all points y in F.

Clearly, every completely regular space is regular, as shown by the argument used to prove that a metric space is regular. Constructing an example of a regular space that is not completely regular takes some work. We will not do this because such spaces are atypical of the kind most mathematicians will encounter, but the interested reader can see [9] or Example 3 on page 154 of [4].

Many define a space to be completely regular if there is a continuous function $h : X \to [0, 1]$ such that $h(x) = 1$ and $h(y) = 0$ for all y in F. This is equivalent to the definition. In fact, suppose that $f : X \to \mathbb{R}$ is as in the preceding definition. If we put $g = f \vee 0$ and then $h = g \wedge 1$, we see that $h : X \to [0, 1]$ is a continuous function (Proposition 2.3.11) with the required properties. Let us record this.

Proposition 3.2.6. *A topological space X is completely regular if and only if for any closed subset F and any point x in $X\backslash F$ there is a continuous function $f : X \to [0, 1]$ such that $f(x) = 1$ and $f(y) = 0$ for all y in F.*

Example 3.2.7. (a) If X is a topological space such that for every closed set F and every point x in $X\backslash F$ there is a function $g : X \to \mathbb{R}$ such that $g(y) = 0$ for all y in F and $g(x) \neq 0$, then X is completely regular. In fact, if such a function exists, then $f(y) = g(x)^{-1}g(y)$ shows that X satisfies the definition of completely regular.

(b) Every metric space (X, d) is completely regular. This follows from (a) since $g(y) = \operatorname{dist}(y, F)$ is continuous. Also, Urysohn's Lemma readily proves this, as we saw in Proposition 3.2.4.

Proposition 3.2.8. *If X is completely regular and Y is a subset of X and has its relative topology from X, then Y is completely regular.*

Proof. Let D be a relatively closed subset of Y, and let $y \in Y \backslash D$. There is a closed subset F of X such that $F \cap Y = D$; so $y \notin F$. Thus, there is a continuous function $f : X \to \mathbb{R}$ with $f(y) = 1$ and $f(z) = 0$ for all z in F. If $g = f|Y$, then $g \in C(Y)$, $g(y) = 1$, and $g(z) = 0$ for all z in D. Therefore, Y is completely regular. ∎

Completely regular spaces guarantee the existence of many nonconstant continuous functions. This is underlined by the next theorem. Recall the definition of the weak topology on a set defined by a collection of functions (Proposition 2.6.2). We will need the following lemma.

Lemma 3.2.9. *If $\mathcal{T}$ is the weak topology defined on X by the collection of functions $\{f_i : X \to X_i : i \in I\}$, then a net $\{x_\alpha\}$ in X converges to x in $(X, \mathcal{T})$ if and only if $f_i(x_\alpha) \to f_i(x)$ for all i in I.*

Proof. Since each function $f_i : (X, \mathcal{T}) \to X_i$ is continuous, the proof of half the corollary is immediate. For the other half assume that $f_i(x_\alpha) \to f_i(x)$ for all i in I. If $G \in \mathcal{T}$ and $x \in G$, then the definition of a subbase implies there are $i_1, \ldots, i_n$ in I and open sets U_{i_k} in X_{i_k} for $1 \leq k \leq n$ such that $x \in \bigcap_{k=1}^n f_{i_k}^{-1}(U_{i_k}) \subseteq G$. Since $f_{i_k}(x_\alpha) \to f_{i_k}(x)$, there is an α_k with $f_{i_k}(x_\alpha) \in U_{i_k}$ for all $\alpha \geq \alpha_k$. If we let $\alpha_0 \geq \alpha_k$ for $1 \leq k \leq n$, then when $\alpha \geq \alpha_0$, we have that $x_\alpha \in \bigcap_{k=1}^n f_{i_k}^{-1}(U_{i_k}) \subseteq G$. Hence $x_\alpha \to x$ in $(X, \mathcal{T})$. ∎

Theorem 3.2.10. *If X is a topological space, then the following statements are equivalent.*

(a)   *X is completely regular.*
(b)   *The topology on X is the weak topology defined by the functions in $C(X)$.*
(c)   *The topology on X is the weak topology defined by the functions in $C_b(X)$.*
(d)   *A net $\{x_i\}$ in X converges to x if and only if $f(x_i) \to f(x)$ for each continuous function $f : X \to [0, 1]$.*

Proof. Let $\mathcal{T}$ denote the topology on X, $\mathcal{T}_c$ the weak topology on X defined by $C(X)$, and $\mathcal{T}_b$ the weak topology on X defined by $C_b(X)$.

(a) *and* (d) *are equivalent.* If $x_i \to x$, then $f(x_i) \to f(x)$ for every continuous function, even without the assumption of complete regularity. Now assume X is completely regular and $f(x_i) \to f(x)$ for each continuous function $f : X \to [0, 1]$. If G is a neighborhood of x, then Proposition 3.2.6 implies there is a continuous function $f : X \to [0, 1]$ with $f(x) = 1$ and

$f(y) = 0$ for all y in $X \backslash G$. Thus, there is an i_0 such that for $i \geq i_0$ we have that $f(x_i) > \frac{1}{2}$, which says that $x_i \in G$.

(a) *implies* (b). We want to show that when X is completely regular, $\mathcal{T} = \mathcal{T}_c$; this means we want to show that if $\tau : (X, \mathcal{T}) \to (X, \mathcal{T}_c)$ is the identity map, then τ is a homeomorphism. If $\{x_i\}$ is a net in X and $x_i \to x$ $(\mathcal{T})$, then $f(x_i) \to f(x)$ for every f in $C(X)$. By Lemma 3.2.9, this implies $x_i \to x$ $(\mathcal{T}_c)$, so τ is continuous. Conversely, if $x_i \to x$ $(\mathcal{T}_c)$, then $f(x_i) \to f(x)$ for every f in $C(X)$. Since (a) and (d) are equivalent, this implies that $x_i \to x$ $(\mathcal{T})$. Thus, τ is a homeomorphism.

(b) *implies* (c). As in the preceding paragraph, we want to show that the identity map $\tau : (X, \mathcal{T}) \to (X, \mathcal{T}_b)$ is a homeomorphism. Clearly, τ is continuous. Now assume $x_i \to x$ in $(X, \mathcal{T}_b)$. Since (b) holds, we want to show that $f(x_i) \to f(x)$ for every f in $C(X)$. Let (a, b) be an arbitrary bounded open interval in $\mathbb{R}$ such that $a < f(x) < b$. For each y in X, define $g(y) = \min\{f(y), b\}$ and $h(y) = \max\{g(y), a\}$. Since $h \in C_b(X)$ and $h(x) = f(x)$, there is an i_0 such that $a < h(x_i) < b$ when $i > i_0$. Fix $i \geq i_0$. From the definition of h we get that $a < h(x_i) = g(x_i) < b$. Again from the definition of g we have that $a < g(x_i) = f(x_i) < b$. Since (a, b) was arbitrary, we have that $f(x_i) \to f(x)$, completing the proof of this part.

(c) *implies* (d). Since each continuous function from X into the closed unit interval belongs to $C_b(X)$, this is immediate by the lemma.   ∎

So the preceding theorem says, in vivid terms, that the topology of a completely regular space is determined by its real-valued continuous functions.

Theorem 3.2.11. *If $\{X_i : i \in I\}$ is a collection of topological spaces and $X = \prod_i X_i$, then X is completely regular if and only if each X_i is completely regular.*

Proof. First assume that X is completely regular, fix a j in I, let F_j be a closed subset of X_j, and let $x_j \in X_j \backslash F_j$. If x is any point in X whose jth coordinate is x_j, then $x \notin \pi_j^{-1}(F_j)$, and this latter set is closed. Let $f : X \to \mathbb{R}$ be a continuous function with $f(x) = 1$ and $f(y) = 0$ for all y in $\pi_j^{-1}(F)$. Now define the map from $X_j \to X$ by $z \mapsto \{z_i\}$, where $z_j = z$ and $z_i = x_i$ for all $i \neq j$. Using nets it follows that this map is continuous. (Verify!) Thus $f_j : X_j \to \mathbb{R}$ defined by $f_j(z) = f(\{z_i\})$ is the composition of two continuous functions and is therefore continuous. Also, $f_j(x_j) = 1$ and $f_j(y_j) = 0$ for all y_j in F_j. Therefore, X_j is completely regular.

Now assume each X_i is completely regular, let F be a closed subset of X, and let $x \in X \backslash F$. From the definition of the product topology this implies there are indices $i_1, \ldots, i_n$ in I and, for $1 \leq k \leq n$, there is a neighborhood G_{i_k} of the point x_{i_k} such that $x \in \bigcap_{k=1}^{n} \pi_{i_k}^{-1}(G_{i_k}) \subseteq X \backslash F$. For $1 \leq k \leq n$, let $f_{i_k} : X_{i_k} \to \mathbb{R}$ such that $f_{i_k}(x_{i_k}) = 1$ and $f_{i_k}(y_{i_k}) = 0$ for all y_{i_k} in $X_{i_k} \backslash G_{i_k}$. Define $f : X \to \mathbb{R}$ by $f(\{y_i\}) = f_{i_1}(y_{i_1}) \cdots f_{i_n}(y_{i_n})$. This function f is continuous (Exercise 12), $f(x) = 1$, and $f(y) = 0$ for y in F.   ∎

Exercises

(1) If X is the real line with the topology generated by the subbase consisting of all the open intervals and the set $\mathbb{Q}$, show that X is Hausdorff but not regular.

(2) Prove that if X is a regular topological space and $E \subseteq X$, then E with its relative topology is regular.

(3) Suppose $X = F_1 \cup F_2$, where F_1 and F_2 are closed, and $F_1 \cap F_2 = \emptyset$. Show that if both F_1 and F_2 with their relative topologies are regular, then X is regular.

(4) If X is a topological space and for each point x in X there is an open set G such that $x \in G$ and $\operatorname{cl} G$ with its relative topology is a regular space, then X is regular.

(5) Show that X is regular if and only if for any two distinct points x and y there are open sets U and V such that $x \in U, y \in V$, and $\operatorname{cl} U \cap \operatorname{cl} V = \emptyset$.

(6) If X is regular and A is a closed subset of X, show that

$$A = \bigcap \{U : U \text{ is open and } A \subseteq U\}.$$

Is the converse true?

(7) Complete the proof of Proposition 3.2.3(b). (Hint: use Exercise 2.6.1.)

(8) As in Example 2.8.2(e), for a topological space X and a subset F define the equivalence relation on X by $x \sim y$ when $x = y$ or when $x, y \in F$. Denote the resulting quotient space by X/F. (a) Show that if F is a closed set and X is regular, then X/F is Hausdorff. (b) Show that if X/F is Hausdorff for every closed subset F of X, then X is regular.

(9) Let X be a topological space, and define a relation $\sim$ on X by declaring that $x \sim y$ if $f(x) = f(y)$ for every continuous function $f : X \to [0, 1]$. (a) Show that $\sim$ is an equivalence relation on X. (b) Show that $X/\sim$ is completely regular.

(10) If X is a connected completely regular space that is not a singleton, show that X has uncountably many points.

(11) Say that a topological space Z has Property P if for any two distinct points z, w in Z there is a continuous function $f : Z \to \mathbb{R}$ such that $f(z) \neq f(w)$. Show that X is completely regular if and only if for every closed subset F of X the quotient space X/F (Exercise 8) has Property P.

(12) Show that the function $f : X \to \mathbb{R}$ defined in the second half of the proof of Theorem 3.2.11 is continuous.

3.3. Normal Spaces

This section continues the last one with the final separation property. The reader should be alert because this section contains a disproportionately large number of significant results.

Definition 3.3.1. A topological space X is *normal* if for any pair of disjoint closed subsets A and B there are disjoint open sets U and V such that $A \subseteq U$ and $B \subseteq V$.

Unlike normal subgroups in algebra, normal topological spaces are the usual occurrence in most parts of mathematics. Exceptions can be found, but perhaps the title "normal" is deserved for these topological spaces. It is immediate that a normal space is regular because every singleton set is a closed set. It is far from clear that a normal space is completely regular, but this follows from one of the deeper results on normal spaces, an extension of Urysohn's Lemma to this setting. (See Theorem 3.3.4 below.)

Proposition 3.3.2. *If X is a topological space, then the following statements are equivalent.*

(a) X *is normal.*

(b)   *If A is a closed set and G is an open set with $A \subseteq G$, then there is an open set U with $A \subseteq U \subseteq \operatorname{cl} U \subseteq G$.*

(c)   *If A and B are disjoint closed sets, then there is an open set V such that $B \subseteq V$ and $A \cap \operatorname{cl} V = \emptyset$.*

Proof. Assume (a) and let A and G be as in (b). Let $B = X \backslash G$, and apply the definition of normality to find disjoint open sets U and V with $A \subseteq U, B \subseteq V$. Thus, $\operatorname{cl} U \subseteq X \backslash V \subseteq X \backslash B = G$. Now assume (b) and let A and B be as in (c). If $G = X \backslash B$, then G is open and $A \subseteq G$. Thus, there is an open set U with $A \subseteq U \subseteq \operatorname{cl} U \subseteq X \backslash B$. Put $V = X \backslash \operatorname{cl} U$; clearly, $B \subseteq V$. Since $V \subseteq X \backslash U$, we have that $\operatorname{cl} V \subseteq X \backslash U$, and so $A \cap \operatorname{cl} V = \emptyset$. Now assume that (c) holds and A and B are disjoint closed sets. If V is as in (c) and $U = X \backslash \operatorname{cl} V$, then U and V are disjoint open sets, $A \subseteq U$, and $B \subseteq V$. Therefore, X is normal. ∎

Proposition 3.3.3. (a)   *Every metric space is normal.*

(b)   *Every compact space is normal.*

(c)   *If X is normal and F is a closed subset of X, then F with its relative topology is normal.*

Proof.   (a) This is an easy application of Theorem 1.3.9.

(b) By Proposition 3.2.4, X is regular. Thus, if A and B are disjoint closed sets, then for every point a in A there are disjoint open sets U_a and V_a such that $a \in U_a$ and $B \subseteq V_a$. Now $\{U_a : a \in A\}$ is an open cover of A, which, being a closed subset of a compact space, is compact. Thus, there are points $a_1, \ldots, a_n$ with $A \subseteq \bigcup_{k=1}^{n} U_{a_k} \equiv U$; put $V = \bigcap_{k=1}^{n} V_{a_k}$. It follows that U and V are disjoint open sets, $A \subseteq U$, and $B \subseteq V$.

(c) Exercise 2.
∎

It is not true that every subspace of a normal space is normal, nor is it the case that the product of even two normal spaces is normal. The reader can consult Example 3 on page 144 of [4] for two normal spaces whose product is not normal. (This reference also contains an example of a regular space that is not normal.) Finding an example of a subspace of a normal space that is not normal requires additional effort. See Example 4 on page 145 of [4], where an open subset of a compact space is shown not to be normal. (This uses spaces of ordinal numbers, which is discussed subsequently in § 3.6.) We might mention that every subspace of a normal space is, however, completely regular. (See Exercise 7.)

Recall Urysohn's Lemma proved for metric spaces (Theorem 1.3.9). Here we get the same conclusion for normal spaces, though the proof is more difficult. Before stating and proving this result we need some facts about dyadic rational numbers. Let $D_0 = \{0, 1\}$, and for $n \geq 1$ let $D_n = \{\frac{a}{2^n} : a \in \mathbb{N}, a$ is odd, and $0 < a < 2^n\}$. Put $D = \bigcup_{n=0}^{\infty} D_n$; this is the set of *dyadic rational numbers* in the closed unit interval. This set has many interesting properties, including the fact that it is dense in the unit interval (Exercise 5). There is more on dyadic rational numbers in § A.5.

Theorem 3.3.4 (Urysohn's Lemma). *If X is normal and A and B are disjoint closed subsets, then there is a continuous function $f : X \to [0, 1]$ such that $f(a) = 1$ for all a in A and $f(b) = 0$ for all b in B.*

Proof. We let $G = X \backslash B$; so $A \subseteq G$. The strategy of the proof is as follows: if D is the set of dyadic rational numbers in $[0, 1]$, then for every t in D we will construct a subset U_t of X satisfying the following:

3.3.5
$$\begin{cases} \text{(i)} & U_t \text{ is open when } t < 1 \\ \text{(ii)} & \text{cl } U_t \subseteq U_s \text{ for } s < t. \end{cases}$$

We will then use this family to construct the continuous function, but first let us get the sets. For this we use induction where of each $n \geq 0$ we will construct the sets $\{U_t\}$ when t belongs to the set D_n defined before the proof.

We start with $n = 0$ and the set D_0. Let $U_1 = A$ and $U_0 = G = X \backslash B$. [Given that U_1 is closed, we note that (3.3.5)(i) is not violated.] When $n = 1$, we have $D_1 = \{\frac{1}{2}\}$. Since X is normal we can find an open set $U_{\frac{1}{2}}$ with $\text{cl } U_1 = A \subseteq U_{\frac{1}{2}} \subseteq \text{cl } U_{\frac{1}{2}} \subseteq U_0$. Let $n \geq 2$, and assume we have constructed the sets U_t for t in $\bigcup_{k=0}^{n-1} D_k$ and that they satisfy (3.3.5). Let $t = \frac{a}{2^n} \in D_n$, with a odd. Since a is an odd integer, $\alpha = \frac{a-1}{2^n}, \beta = \frac{a+1}{2^n} \in \bigcup_{k=0}^{n-1} D_k$, and so U_α and U_β are already defined with $\text{cl } U_\beta \subseteq U_\alpha$. Since X is normal, there is an open set U_t with $\text{cl } U_\beta \subseteq U_t \subseteq \text{cl } U_t \subseteq U_\alpha$. In this way, we define U_t satisfying (3.3.5) for all t in $\bigcup_{k=0}^{n} D_k$. By mathematical induction we have the sets U_t for all dyadic rational numbers.

Now to define f. When $x \in X \backslash U_0 = B$, let $f(x) = 0$; when $x \in U_0 = G = X \backslash B$, define $f(x) = \sup\{t \in D : x \in U_t\}$. It is clear that $f(x) = 1$ for x in $A = U_1$ and $0 \leq f(x) \leq 1$ for all x in X. What must be proven is that f is continuous. For this let $0 \leq \alpha < \beta \leq 1$. Note that $\alpha < f(x)$ if and only if $x \in U_t$ for some $t > \alpha$; thus,

$$f^{-1}((\alpha, 1]) = \bigcup \{U_t : \alpha < t\},$$

an open set. Also, $f(x) \geq \beta$ if and only if $x \in U_t$ for every $t < \beta$; thus,

$$f^{-1}([\beta, 1]) = \bigcap \{U_t : t < \beta\} = \bigcap \{\operatorname{cl} U_s : s < \beta\},$$

a closed set. Hence, $f^{-1}([0, \beta))$ is open, so that $f^{-1}((\alpha, \beta)) = f^{-1}((\alpha, 1]) \cap f^{-1}([0, \beta))$ is open. Therefore, f is continuous. (Why?) ∎

Corollary 3.3.6. *If X is normal, A and B are disjoint closed subsets, and α and β are real numbers with $\alpha < \beta$, then there is a continuous function $f : X \to [\alpha, \beta]$ such that $f(a) = \alpha$ for all a in A and $f(b) = \beta$ for all b in B.*

Proof. Note that $\tau(t) = t\beta + (1 - t)\alpha$ is a homeomorphism of $[0, 1]$ onto $[\alpha, \beta]$, with $\tau(0) = \alpha, \tau(1) = \beta$. According to Urysohn's Lemma there is a continuous function $g : X \to [0, 1]$ with $g(a) = 0$ for all a in A and $g(b) = 1$ for all b in B. The function $f = \tau \circ g$ is the desired function. ∎

Corollary 3.3.7. *Every normal space is completely regular.*

Next we show that normal spaces enjoy a special property—continuous functions defined on a closed subset can be extended to the entire space. It is a deep property, and its establishment starts with a lemma.

Lemma 3.3.8. *If X is a normal space, C is a closed subset of X, and $f : C \to \mathbb{R}$ is a continuous function with $|f(c)| \leq \gamma$ for all c in C, then there is a continuous function $g : X \to \mathbb{R}$ satisfying the following for all x in X and all c in C:*

(i) $|g(x)| \leq \gamma/3$

(ii) $|f(c) - g(c)| \leq 2\gamma/3.$

Proof. Let $A = \{c \in C : f(c) \leq -\gamma/3\}, B = \{c \in C : f(c) \geq \gamma/3\}$. Since f is continuous on the closed set C, A and B are disjoint closed subsets of X. By Corollary 3.3.6, there is a continuous function $g : X \to [-\gamma/3, \gamma/3]$ with $g(a) = -\gamma/3$ for all a in A and $g(b) = \gamma/3$ for all b in B. So g satisfies (i). We show that it also satisfies (ii) on a case-by-case basis. Let $c \in C$. First note that if $c \in A$, then $-\gamma \leq f(c) \leq -\gamma/3$ and $g(c) = -\gamma/3$, so (ii) holds. Similarly, (ii) holds for c in B. If $c \notin (A \cup B)$, then $|f(c)| < \gamma/3$; since $|g(c)| \leq \gamma/3$, (ii) is also valid in this case. ∎

Theorem 3.3.9 (Tietze's[2] Extension Theorem). *If X is a normal topological space, C is a closed subset of X, and $f : C \to [\alpha, \beta]$ is a continuous function,*

---

[2]Heinrich Franz Friedrich Tietze was born in 1880 in Schleinz, Austria. In 1898 he entered the Technische Hochschule in Vienna. He continued his studies in Vienna and received his doctorate in 1904 and his habilitation in 1908 with a thesis in topology. His academic career was interrupted by service in the Austrian army in World War I; just before this he had obtained the present theorem. After the war he held a position first at Erlangen and then at Munich, where he remained until his retirement. In addition to this theorem, he made other contributions to topology and did significant work in combinatorial group theory, a field in which he was one of the pioneers. He had 12 Ph.D. students, all at Munich, where he died in 1964.

then there is a continuous function $F : X \to [\alpha, \beta]$ such that $F(c) = f(c)$ for every c in C.

Proof. Using the map τ in the proof of Corollary 3.3.6, it suffices to assume that $f : C \to [0, 1]$. Applying the preceding lemma we can find a continuous function $g_0 : X \to [-3^{-1}, 3^{-1}]$ such that $|f(c) - g_0(c)| \leq 2/3$ for all c in C. Now apply the lemma again to the function $(f - g_0) : C \to [-2/3, 2/3]$ to obtain a continuous function $g_1 : X \to [-3^{-1}(2/3), 3^{-1}(2/3)]$ such that $|(f - g_0)(c) - g_1(c)| \leq (2/3)^2$. Continuing this line of reasoning and applying induction, we can establish the following claim.

Claim. For each integer $n \geq 1$ there is a continuous function $g_n : X \to \mathbb{R}$ satisfying the following: (i) $|g_n(x)| \leq 3^{-1}(2/3)^n$ for all x in X; (ii) $|f(c) - \sum_{k=0}^{n} g_k(c)| \leq (2/3)^{n+1}$ for all c in C.

The details of the induction argument required to establish this claim are left to the reader (Exercise 8).

Because of (i) in the claim, the Weierstrass M-test (Exercise 3.1.2) implies $g = \sum_{n=0}^{\infty} g_n$ converges in $C_b(X)$; so $g \in C_b(X)$ and, summing the appropriate geometric series, for every x in X we have $|g(x)| \leq 1$. Also (ii) in the claim implies that g is an extension of f. The only thing lacking is that g may take on negative values. If, however, we put $F = g \vee 0$; then F is continuous (Proposition 2.3.11), $F(x) \in [0, 1]$ for all x, and, because f is nonnegative, F is an extension of f.   ∎

A fact worth emphasizing in the Tietze Extension Theorem is that the image of the extension F is contained in the same interval that contains the image of f. This is not to say, however, that the functions f and F have the same image. (Can you furnish an example where the ranges differ? You might find one with $X = \mathbb{R}$.)

The next result is an important application of Urysohn's Lemma.

Theorem 3.3.10 (Partition of Unity). *If X is normal and $\{G_1, \ldots, G_n\}$ is an open cover of X, then there are continuous functions $\phi_1, \ldots, \phi_n$ from X into $\mathbb{R}$ with the following properties:*

(a) $0 \leq \phi_k(x) \leq 1$ for $1 \leq k \leq n$;

(b) $\phi_k(x) = 0$ when $x \notin G_k$ and $1 \leq k \leq n$;

(c) $\sum_{k=1}^{n} \phi_k(x) = 1$ for all x in X.

Proof. We prove this by induction. First assume $n = 2$: $X = G_1 \cup G_2$. $X \backslash G_2$ is a closed set that is contained in G_1 and so Urysohn's Lemma implies there is a continuous function $\phi_1 : X \to [0, 1]$ with $\phi_1(x) = 1$ on $X \backslash G_2$ and 0 on $X \backslash G_1$. Put $\phi_2 = 1 - \phi_1$. It is routine to check that the three conditions are satisfied. Now assume the theorem is true for some $n \geq 2$, and let $\{G_1, \ldots, G_n, G_{n+1}\}$ be an open cover of X. The induction hypothesis applied to the open cover $\{G_1, \ldots, G_{n-1}, G_n \cup G_{n+1}\}$ implies the existence of continuous functions $\psi_k : X \to [0, 1]$ for $1 \leq k \leq n$ such that $\sum_{k=1}^{n} \psi_k = 1$,

$\psi_k(x) = 0$ for $x \notin G_k$ and $1 \leq k \leq n-1$, and $\psi_n(x) = 0$ when $x \notin G_n \cup G_{n+1}$. Now consider the open cover of X consisting of the two sets $G_1 \cup \cdots \cup G_n$ and G_{n+1}, and find continuous functions θ_1, θ_2 with $\theta_1 + \theta_2 = 1$, $\theta_1(x) = 0$ when $x \notin G_1 \cup \cdots G_n$, and $\theta_2(x) = 0$ when $x \notin G_{n+1}$. Put $\phi_k = \theta_1 \psi_k$ when $1 \leq k \leq n$ and $\phi_{n+1} = \theta_2 \psi_n$. Clearly, each of these functions has its range in the closed unit interval, so (a) is satisfied. Also $\sum_{k=1}^{n+1} \phi_k = \theta_1 \sum_{k=1}^{n} \psi_k + \theta_2 \psi_n = 1$, so (c) holds. If $1 \leq k \leq n-1$ and $\phi_k(x) > 0$, then $\psi_k(x) > 0$; so it must be that $x \in G_k$. If $\phi_n(x) > 0$, then $\theta_1(x) > 0$ and $\psi_n(x) > 0$. Hence $x \in G_1 \cup \cdots \cup G_n$ and $x \in G_n \cup G_{n+1}$; thus, $x \in G_n$. If $\phi_{n+1}(x) > 0$, then $\theta_2(x) > 0$, and so $x \in G_{n+1}$. By induction, this completes the proof. ∎

For the open cover $\{G_1, \ldots, G_n\}$ and the functions $\phi_1, \ldots, \phi_n$ as in the preceding theorem, we say that these functions are a *partition of unity subordinate to the cover*. The reason this result is called a partition of unity is that it divides the constantly 1-function into parts that reside inside the open sets G_k. Later (Theorem 3.7.17) we will see a more sophisticated partition of unity theorem.

Corollary 3.3.11. *If K is a closed subset of the normal space X and $\{G_1, \ldots, G_n\}$ are open sets in X that cover K, then there are continuous functions $\phi_1, \ldots, \phi_n$ on X with the following properties:*

(a) *for $1 \leq k \leq n$ and all x in X, $0 \leq \phi_k(x) \leq 1$;*
(b) *for $1 \leq k \leq n$, $\phi_k(x) = 0$ when $x \notin G_k$;*
(c) $\sum_{k=1}^{n} \phi_k(x) = 1$ *for all x in K;*
(d) $\sum_{k=1}^{n} \phi_k(x) \leq 1$ *for all x in X.*

Proof. Note that if we put $G_{n+1} = X \backslash K$, then $\{G_1, \ldots, G_{n+1}\}$ is an open cover of X. Let $\{\phi_1, \ldots, \phi_{n+1}\}$ be a partition of unity subordinate to this cover. The reader can check that the functions $\phi_1, \ldots, \phi_n$ satisfy (a), (b), and (c) since $\phi_{n+1}(x) = 0$ for x in K. For any x in X, $\sum_{k=1}^{n} \phi_k(x) = 1 - \phi_{n+1}(x) \leq 1$, giving (d). ∎

Exercises

(1) Supply the missing details in the proof of Proposition 3.3.2.
(2) Prove Proposition 3.3.3(c).
(3) Show that if X is normal and $f : X \to Y$ is a continuous surjection that is also a closed map, then Y is normal. (A mapping is *closed* if the image of every closed set is closed.)
(4) For disjoint subsets A and B of X and x, y in X, define $x \sim y$ to mean that one of the following holds: $x = y$, $x, y \in A$, $x, y \in B$. (See Exercise 3.2.8 for a related exercise.) (a) Verify that this is an equivalence relation on X. Denote the resulting quotient space by $X/\{A, B\}$. (b) If X is normal and A and B are disjoint closed subsets, show that the

quotient topology on $X/\{A, B\}$ is Hausdorff. (c) If X is a topological space such that $X/\{A, B\}$ is Hausdorff for every pair of disjoint closed subsets A, B, then X is normal.

(5) Show that the dyadic rational numbers are dense in the unit interval.

(6) If X is normal, use Urysohn's Lemma to show that if F is a closed set and G is an open set with $F \subseteq G$, then there is a continuous function $f : X \to [0, 1]$ such that $f(x) = 1$ when $x \in F$ and $\{x \in X : f(x) > 0\} \subseteq G$. (Also see Exercise 11, parts (d) and (e).)

(7) Use Urysohn's Lemma to show that every subspace of a normal space is completely regular.

(8) Use induction to establish the claim in the proof of Tietze's Extension Theorem.

(9) (a) Show that $\phi(x) = \arctan x$ defines a homeomorphism between $\mathbb{R}$ and $(-\pi/2, \pi/2)$. (b) Prove the following extension of Tietze's Extension Theorem. If X is normal, A is a closed subset of X, and $f : A \to \mathbb{R}$ is a continuous function, then there is a continuous function $F : X \to \mathbb{R}$ with $F(a) = f(a)$ for all a in A.

(10) Prove the following extension of Tietze's Extension Theorem. If X is normal, A is a closed subset of X, and $f : A \to \mathbb{R}^q$ is a continuous function, then there is a continuous function $F : X \to \mathbb{R}^q$ with $F(a) = f(a)$ for all a in A. (Hint: use Exercise 9.)

(11) A subset A of a topological space X is called a G_δ-*set* if it is the intersection of a sequence of open sets. A is called an F_σ-*set* if it is the union of a sequence of closed sets. (a) Show that A is a G_δ-set if and only if $X \backslash A$ is an F_σ-set. (b) Show that the set of irrational numbers is a G_δ set. (c) In a metric space show that every closed set is a G_δ-set and every open set is an F_σ-set. (d) In Exercise 6 show that the function f can be chosen with $F = \{x \in X : f(x) = 1\}$ if and only if F is a G_δ-set. (e) In Exercise 6 show that the function f can be chosen with $G = \{x \in X : f(x) > 0\}$ if and only if G is an F_σ-set. (f) Is $\mathbb{Q}$ a G_δ-set in $\mathbb{R}$?

3.4. The Stone–Čech Compactification*

The main result of this section asserts that each completely regular space X is densely contained in a compact space βX in such a way that each bounded continuous function $f : X \to \mathbb{R}$ has a continuous extension $f^\beta : \beta X \to \mathbb{R}$. The use of the word *contained* in this last statement is, strictly speaking, not accurate. The actual statement is that there is a homeomorphism τ from X onto a dense subset of βX such that when $f \in C_b(X)$, $f \circ \tau^{-1}$ has a continuous extension to βX. We are justified in saying that X is contained in βX by the fact that this compact space is unique up to a homeomorphism having special properties as listed in the theorem. Another phrasing of this idea of containment that is used is to say that X is densely embedded in βX.

Theorem 3.4.1 (The Stone[3]–Čech[4] Compactification). *If X is completely regular, then there is a compact space βX and a homeomorphism τ from X onto a dense subset of βX such that for every bounded continuous function f : $X \to \mathbb{R}$ there is a continuous function f^β : $\beta X \to \mathbb{R}$ with $f^\beta \circ \tau = f$. Moreover, βX is unique in the sense that if Z is a compact space with a homeomorphism σ of X onto a dense subset of Z such that for each bounded continuous function f : $X \to \mathbb{R}$ there is a continuous f^Z : $Z \to \mathbb{R}$ with $f^Z \circ \sigma = f$, then Z is homeomorphic to βX.*

The proof of this theorem is involved. Before starting the proof, let us ruminate on it. If we identify X with its homeomorphic image in βX, then the result asserts that each completely regular space X is densely *contained* in a compact space βX in such a way that each bounded continuous function f : $X \to \mathbb{R}$ has a continuous extension f^β : $\beta X \to \mathbb{R}$. The use of the word *contained* in this last statement is amply justified by the fact that the compact

---

[3]Marshall H. Stone was born in 1902 in New York. His father was Harlan Stone who, after time as the dean of the Columbia Law School, became a member of the US Supreme Court, including a term as its Chief Justice. Marshall Stone entered Harvard in 1919 intending to study law. He soon switched to mathematics and received his doctorate in 1926 under the direction of David Birkhoff. Though he had brief appointments at Columbia and Yale, most of his early career was spent at Harvard. His initial work continued the direction it took under Birkhoff, but in 1929 he started working on hermitian operators. His American Mathematical Society book *Linear Transformations in Hilbert space and Their Applications to Analysis* became a classic. Indeed, a read of that book today shows how the arguments and clarity would easily lead to the conclusion that it is a contemporary monograph. During World War II he worked for the Navy and the War Department, and in 1946 he left Harvard to become the chair of the mathematics department at the University of Chicago. He said that this decision was arrived at because of "my conviction that the time was also ripe for a fundamental revision of graduate and undergraduate mathematical education." Indeed he transformed the department at Chicago. The number of theorems that bear his name is impressive. Besides the present theorem there is the Stone–Weierstrass Theorem, the Stone–von Neumann Theorem, the Stone Representation Theorem in Boolean algebras, and Stone's Theorem on one-parameter semigroups. He stepped down as chair at Chicago in 1952 and retired in 1968, but then went to the University of Massachusetts, where he taught in various capacities until 1980. He loved to travel, and on a trip to India in 1989, he died in Madras. He had 14 doctoral students, including Richard Kadison and George Mackey.

[4]Eduard Čech was born in 1893 in Stracov, Bohemia, which is now part of the Czech Republic. His father was a policeman and he was his parents' fourth child. He quickly displayed mathematical talent and decided he wanted to be a school teacher. With that aim he entered Charles University in Prague in 1912. In 1915 he was drafted into the Austro-Hungarian army; after the war he returned to the university, obtained his degree, began teaching at a secondary school in Prague, and continued his research. This led to a doctorate in 1920. He worked on projective differential geometry with a paper that appeared in 1921, and this secured for him a scholarship to study with Fubini in Turin. They collaborated on a two-volume monograph on the subject. He received his habilitation in 1922 after leaving Italy. In 1922 he was made chair of mathematics at a new university in Brno; the next year he was appointed Extraordinary Professor. Around this time he began to work on topology, the source of his most famous work that includes Čech cohomology; as part of the process he started a topology seminar at Brno. It was in 1937 that he obtained the present result, with Stone proving it in the same year. After the takeover of Czechoslovakia by the Nazis, the universities were closed. However, the seminar continued to meet at the home of one of his students, Pospisl, until he was arrested by the Gestapo in 1941. (Pospisl was released in 1944 but died shortly thereafter.) At the end of the war Čech moved to Charles University in Prague and became involved in administration. He was the Director of the Mathematical Research Institute of the Czech Academy of Sciences in 1947, Director of the Central Mathematical Institute in 1950, and Director of the Czech Academy in 1952. After a rather long hiatus in his publishing, extending through the war, he resumed it in the 1950s with an interest in differential geometry. In 1956 he was appointed First Director of the Mathematical Institute of the Charles University of Prague. His health, however, began to fade, and he died in 1960 in Prague.

space βX is unique up to a homeomorphism having special properties as listed in the theorem. Another phrasing of this idea of containment that can be and is used is to say that X is densely embedded in βX.

The space βX is called the *Stone–Čech compactification* of X. This compactification can be very complicated. We will see some exercises below that show this, but consider one example: $X = (0, 1]$. You might be tempted to think that the Stone–Čech compactification of this set is the closed unit interval since X is densely contained in it. This is not the case. For example, if we let $f : (0, 1] \to \mathbb{R}$ be the bounded continuous function $f(t) = \sin t^{-1}$, then f must have a continuous extension to $\beta(0, 1]$, but there is no such extension to the closed interval (Exercise 3). So if we take a compact space Z and let X be a dense subset of Z, then X is completely regular (Proposition 3.2.8), but Z may fail miserably to be βX. Giving a specific representation of βX for the most innocent noncompact spaces X is usually impossible.

Proof. *Existence.* Let $\mathcal{F}$ denote all the continuous functions from X into $[0, 1]$, and for each f in $\mathcal{F}$ let $X_f = [0, 1]$, a copy of the closed unit interval. Let $\Omega = \prod\{X_f : f \in \mathcal{F}\}$, and define $\tau : X \to \Omega$ by $\tau(x) = \{f(x) : f \in \mathcal{F}\}$; that is, the coordinate of $\tau(x)$ corresponding to each f in $\mathcal{F}$ is given by $\tau(x)_f = f(x)$. Let $\beta X = \mathrm{cl}\,[\tau(X)]$.

Claim. $\tau : X \to \tau(X)$ is a homeomorphism.

In fact, τ is injective since, if x and y are two distinct points in X, then Proposition 3.2.6 implies there is an f in $\mathcal{F}$ with $f(x) = 1$ and $f(y) = 0$; so $\tau(x) \neq \tau(y)$. τ is surjective by definition. If $\{x_i\}$ is a net in X and $x \in X$, then an application of Theorem 3.2.10 shows that $x_i \to x$ if and only if $\tau(x_i)_f \to \tau(x)_f$ for every f in $\mathcal{F}$. By Exercise 2.7.7, this says that $x_i \to x$ if and only if $\tau(x_i) \to \tau(x)$ in the product space Ω. Hence we have established the claim.

Claim. If $f \in C_b(X)$, then there is an f^β in $C(\beta X)$ such that $f^\beta \circ \tau$.

First note that if $f \in \mathcal{F}$ and we define f^β to be the restriction of the projection map π_f to βX, then f^β has the desired property. Now if $f \in C_b(X)$ and $a < f(x) < b$ for all x in X, then $g = (b - a)^{-1}(f - a) \in \mathcal{F}$. Therefore, g^β exists, and it is a simple matter to verify that defining $f^\beta = (b - a)g^\beta + a$ has all the properties we desire. (Verify!)

Uniqueness. Let Z and σ be as in the statement of the theorem, and let Ω be the product space in the proof of existence. Define $\omega : Z \to \Omega$ by $\omega(z) = \{f^Z(z) : f \in \mathcal{F}\}$. Note that for x in X and f in $\mathcal{F}$, $\omega(\sigma(x))_f = f(x) = \tau(x)_f$. Hence, $f^Z[\omega(\sigma(x))] = f(x)$ for every x in X, $\omega(Z) \subseteq \beta X$, and

$\tau = \omega \circ \sigma$. This results in the following commutative mapping diagram.

3.4.2

The meaning of the term *commutative* here is that if we begin with a point x in the space X, no matter which path of maps in this diagram we follow to reach βX or $\mathbb{R}$, we obtain the same result. The fact that the diagram is commutative was established by the equations given prior to the diagram. Commutative diagrams are frequently used in topology. The purpose of presenting this one is to help the reader keep track of the mappings and spaces.

Claim. $\omega : Z \to \omega(Z)$ is a homeomorphism.

First we note that ω is continuous since for every f in $\mathcal{F}$, $\pi_f \circ \omega = f^Z$, and we can invoke Proposition 2.6.5. Let z and w be distinct points in Z; because compact spaces are normal, there is a continuous function $h : Z \to [0, 1]$ such that $h(z) = 0$ and $h(w) = 1$. If $f = h \circ \sigma$, then $f \in \mathcal{F}$. Since $\sigma(X)$ is dense in Z, it must be that $h = f^Z$. Thus, $\omega(z)_f = f^Z(z) = 0 \neq 1 = f^Z(w) = \omega(w)_f$. Thus, $\omega(z) \neq \omega(w)$, and ω is injective. So ω is an injective continuous mapping onto its image with a compact domain; by Exercise 2.4.3, the claim is established.

Since $\omega \circ \sigma = \tau$ on X, $\tau(X) \subseteq \omega(Z)$. Therefore, $\beta X \subseteq \omega(Z)$ because $\omega(Z)$ is compact. On the other hand, the image of σ is dense in Z, and this implies that $\omega(Z) \subseteq \beta X$, so that these two compact spaces are equal. This completes the proof. ∎

The notation used in the statement of the theorem is standard. From now on, including in the exercises, we will assume that $X \subseteq \beta X$ and that for each f in $C_b(X)$, f^β is the extension of f to βX.

It is not just real-valued continuous functions that can be extended to the Stone–Čech compactification.

Theorem 3.4.3. *If X is completely regular and Z is a compact space, then every continuous mapping $\eta : X \to Z$ has a continuous extension $\eta^\beta : \beta X \to Z$.*

Proof. If $\xi \in \beta X$, then let $\{x_i\}$ be a net in X such that $x_i \to \xi$. Since Z is compact, the net $\{\eta(x_i)\}$ has a cluster point z in Z. If $h \in C(Z)$, then $h(\eta(x_i)) \to_{cl} h(z)$. But $h \circ \eta \in C_b(X)$, so it must be that $h \circ \eta(x_i) \to (h \circ \eta)^\beta(\xi)$. Thus, $h(z) = (h \circ \eta)^\beta(\xi)$ for every h in $C(Z)$. Since compact spaces are normal, this implies that z is the unique cluster point of the net $\{\eta(x_i)\}$. By Theorem 2.7.10, this net converges to z. So any net in X that converges in βX is mapped by η to a net that converges in Z. We want to define

3.4.4 $$\eta^\beta(\xi) = \lim_i \eta(x_i).$$

We must be sure, however, that η^β is well defined and does not depend on the choice of net in X that converges to ξ. But if $\{y_j\}$ is a second net in X that converges to ξ and $\eta(y_j) \to w$ in Z, then for every h in $C(Z)$ we have that $h(w) = \lim_j h(\eta(y_j)) = (h \circ \eta)^\beta(\xi) = \lim_i h(\eta(x_i)) = h(z)$. Once again, Urysohn's Lemma implies $w = z$. Therefore, η^β is well defined.

Clearly, η^β is an extension of η, so it remains to show that this extension is continuous. Let us start by recouping from the last paragraph the fact that for each h in $C(Z)$,

$$(h \circ \eta)^\beta = h \circ \eta^\beta.$$

In fact. this follows from (3.4.4): if $\{x_i\}$ is a net in X that converges to ξ, then, by definition, $\eta(x_i) \to \eta^\beta(\xi)$; since h is continuous and $h \circ \eta \in C_b(X)$, $h(\eta^\beta(\xi)) = \lim_i h(\eta(x_i)) = \lim_i (h \circ \eta)(x_i) = (h \circ \eta)^\beta(\xi)$. Hence. if $\{\xi_i\}$ is a net in βX such that $\xi_i \to \xi$ and $h \in C(Z)$,

$$h(\eta^\beta(\xi_i)) = (h \circ \eta)^\beta(\xi_i) \to (h \circ \eta)^\beta(\xi) = h(\eta^\beta(\xi)).$$

But, by Theorem 3.2.10 and the fact that Z is completely regular, this implies that $\eta^\beta(\xi_i) \to \eta^\beta(\xi)$. Therefore, η^β is continuous, and the proof is complete. ∎

Exercises

(1) For a topological space X show that the weak topology defined on X by $C_b(X)$ is the same as the weak topology defined by $\mathcal{F} = \{f \in C_b(X) : 0 \le f \le 1\}$.

(2) If X is a normal space and there is a point ξ in βX such that $\xi = \lim_n x_n$ for some sequence $\{x_n\}$ in X, show that $\xi \in X$. Hence the Stone–Čech compactification is metrizable only if X is already a compact metric space.

(3) Show that if X is completely regular and βX is metrizable, then X is a compact metric space so that $\beta X = X$.

(4) Here we consider the completely regular space $\mathbb{N}$ and its Stone–Čech compactification $\beta \mathbb{N}$. Note that if $E \subseteq \mathbb{N}$, then the characteristic function of E, χ_E (Exercise 1.5.4), is in $C_b(\mathbb{N})$. (a) Show that every component of $\beta \mathbb{N}$ reduces to a single point. (b) Show that $\beta \mathbb{N}$ is uncountable. (Actually, the cardinality of $\beta \mathbb{N}$ is 2^c, where c is the cardinality of $\mathbb{R}$. This is more difficult to prove.)

3.5. Locally Compact Spaces

Here we explore another of the "local" properties that plays a central role in many areas of mathematics.

Definition 3.5.1. A topological space X is *locally compact* if for every point x in X and every neighborhood G of x there is another neighborhood U of x such that $\operatorname{cl} U$ is compact and contained in G.

Locally compact spaces are everywhere and are the most common spaces discussed in analysis. Here are some examples.

Example 3.5.2. (a) Note that a metric space (X, d) is locally compact if and only if for every x in X and every $r > 0$ there is an ϵ with $0 < \epsilon < r$ such that $B(x; \epsilon) \subseteq B(x; r)$ and $\operatorname{cl} B(x; \epsilon)$ is compact.

(b) $\mathbb{R}^q$ is locally compact.

(c) Every compact space is locally compact.

(d) $\mathbb{Q}$ is not locally compact. This follows from Example 1.4.9.

(e) Let Y be any compact space and fix a point y_0 in Y. If $X = Y \setminus \{y_0\}$ has its relative topology, then X is locally compact (Exercise 2).

Proposition 3.5.3. (a) *If X is a topological space, then X is locally compact if and only if for every point x in X there is a neighborhood G of x such that $\operatorname{cl} G$ is compact.*

(b) *If X is a locally compact topological space and $E \subseteq X$ such that E is either open or closed, then E with its relative topology is locally compact.*

(c) *If $\{X_i\}$ is a collection of topological spaces and $X = \prod_i X_i$, then X is locally compact if and only if each X_i is locally compact and all but a finite number of the spaces X_i are compact.*

(d) *If (X, d) is a metric space, then X is locally compact if and only if for every x in X there is an $r > 0$ such that $\operatorname{cl} B(x; r)$ is compact.*

Proof. (a) The proof of one direction in this statement is trivial. For the other direction, let G be as in the statement of (a), and let U be any neighborhood of x; we must show that there is a neighborhood V of x such that $\operatorname{cl} V$ is compact and contained in U. Note that $W = G \cap U$ is a neighborhood of x and $\operatorname{cl} W$ is compact; however, it may not satisfy $\operatorname{cl} W \subseteq U$. For each y in ∂W, let V_y and H_y be disjoint open sets such that $x \in V_y \subseteq W$ and $y \in H_y$. Since $\partial W \subseteq \operatorname{cl} W$, ∂W is compact. Let $y_1, \ldots, y_n \in \partial W$ such that $\partial W \subseteq \bigcup_{j=1}^n H_{y_j}$, and let $V = \bigcap_{j=1}^n V_{y_j}$. Thus, V is a neighborhood of x, $\operatorname{cl} V$ is compact, $\operatorname{cl} V \subseteq X \setminus \bigcup_{j=1}^n H_{y_j}$ since V is a subset of this closed set. Also, $V \subseteq W$ and $V \cap \partial W = \emptyset$, so $\operatorname{cl} V \subseteq W \subseteq U$, as required.

(b) Assume that E is closed and $x \in E$. There is an open subset G in X such that $x \in G$ and $\operatorname{cl} G$ is compact. Thus $U = G \cap E$ is relatively open in E and contains x. Since E is closed, the closure of U in E is contained in $E \cap \operatorname{cl} G$ and hence is compact. It follows from part (a) that E is locally compact. The proof of the case where E is open is Exercise 3.

(c) Suppose that X is locally compact , and fix an index i_0; we will show that X_{i_0} is locally compact. Let x_{i_0} in X_{i_0}, and let x be any point in X with x_{i_0} as its i_0-coordinate. Let G_{i_0} be a neighborhood of x_{i_0}, and consider $G = \pi_{i_0}^{-1}(G_{i_0})$, a neighborhood of x in X. Since X is locally compact, there is a neighborhood U of x such that $\operatorname{cl} U$ is compact and $\operatorname{cl} U \subseteq G$. Thus, $U_{i_0} = \pi_{i_0}(U)$ is open in X_{i_0} (Proposition 2.6.5(a)), U_{i_0} contains x_{i_0}, $\operatorname{cl} U_{i_0}$ is compact since it is contained in $\pi_{i_0}(\operatorname{cl} U)$, and $\operatorname{cl} U_{i_0} \subseteq G_{i_0}$ (Why?). Thus, X_{i_0} is locally compact. Also, note that the definition of the topology on X implies that there is a finite number of indices $i_0, i_1, \ldots, i_n$ such that when $i \neq i_0, 1_i, \ldots, i_n$, $\pi_i(\operatorname{cl} U) = X_i$; hence each of these spaces X_i is compact.

The proof of the converse is Exercise 4.

(d) This is an easy consequence of (a). (Exercise 5.)   ∎

We should note that some restriction on the subset E in part (b) of the preceding proposition is required. For example, we have noted that $\mathbb{Q}$ is not locally compact, even though $\mathbb{R}$ is. Also, see Exercise 1. So there are subsets of a locally compact space that are not locally compact when furnished with their relative topology. Look at Exercise 2 for further information. As a comment on part (d), note that $\mathbb{R}$ with the metric $d(x,y) = |x - y|(1 + |x - y|)^{-1}$ is locally compact since this metric is equivalent to the usual one; however, it is not true that $\operatorname{cl} B(x; r)$ is compact for every value of r (for example, $r = 1$).

The next lemma, which is almost just an observation because of its easy proof, has a use beyond its application in the subsequent theorem.

Lemma 3.5.4. *If X is a locally compact space, K is a compact subset of X, and G is an open set that contains K, then there is an open set U such that $K \subseteq U \subseteq \operatorname{cl} U \subseteq G$ and $\operatorname{cl} U$ is compact.*

Proof. For each x in K, let U_x be a neighborhood such that $\operatorname{cl} U_x$ is compact and contained in G. Since K is compact, there are points $x_1, \ldots, x_n$ in K such that $K \subseteq \bigcup_{k=1}^n U_{x_k}$. If $U = \bigcup_{k=1}^n U_{x_k}$, it has the required properties.   ∎

Theorem 3.5.5. *If X is a locally compact space, K is a compact subset of X, and G is an open set that contains K, then there is a continuous function $f : X \to [0,1]$ such that $f(x) = 1$ for all x in K and $f(x) = 0$ when $x \notin G$.*

Proof. Invoke the preceding lemma to find an open set with $K \subseteq U \subseteq \operatorname{cl} U \subseteq G$ and $\operatorname{cl} U$ compact. Since $\operatorname{cl} U$ is compact, Urysohn's Lemma implies there is a continuous function $g : \operatorname{cl} U \to [0,1]$ with $g(x) = 1$ for x in K and $g(x) = 0$ for x in ∂U. If we define $h : X \backslash U \to \mathbb{R}$ to be the identically 0 function, then we can use Proposition 2.3.4 to get the desired continuous function f.   ∎

There is a temptation to call the preceding theorem Urysohn's Lemma, and some do. After all, if $B = X \backslash G$, then B is a closed set disjoint from K and the function equals 0 on B. So it certainly has the flavor of Urysohn's Lemma. Nevertheless, it adds an extra condition to the hypothesis by requiring that K be compact since we do not know that X is normal. Similarly, there is a temptation to call the Theorem 3.5.7 below Tietze's Extension Theorem, but we will also resist such an impulse for the same reason.

Corollary 3.5.6. *A locally compact space is completely regular.*

Proof. If F is a closed subset of X and $x \in X \backslash F$, put $K = \{x\}$, $G = X \backslash F$, and apply the preceding theorem. ∎

A locally compact space is not necessarily normal. In fact, we pointed out that Example 4 on page 145 of [4] gives an open subset of a compact space that is not normal. By Proposition 3.5.3(b), such an open set with the relative topology is locally compact. (In fact, that example uses the space of ordinal numbers, which is defined in the next section.)

Theorem 3.5.7. *If X is a locally compact space, K is a compact subset, G is an open set with $K \subseteq G$, and $f : K \to [0,1]$ is a continuous function, then there is a continuous function $F : X \to [0,1]$ such that $F(x) = f(x)$ for all x in K and $F(x) = 0$ when $x \notin G$.*

Proof. Again use Lemma 3.5.4 to find an open set U such that $\mathrm{cl}\, U$ is compact and $K \subseteq U \subseteq \mathrm{cl}\, U \subseteq G$. As in the proof of the preceding theorem, $\mathrm{cl}\, U$ is normal, so Tietze's Extension Theorem implies there is a continuous function $F_1 : \mathrm{cl}\, U \to [0,1]$ such that $F_1(x) = f(x)$ for all x in K and $F_1(x) = 0$ when $x \in \partial U$. Now we proceed as in the proof of Theorem 3.5.5 and obtain the sought-after function F. ∎

See Exercise 7.

There is a Baire Category Theorem for locally compact spaces, and this time we succumb to the temptation and call it precisely that.

Theorem 3.5.8 (Baire Category Theorem). *If X is locally compact and $\{U_n\}$ is a sequence of open subsets of X each of which is dense, then $\bigcap_{n=1}^{\infty} U_n$ is dense.*

The proof of this is along the lines of the proof of the Baire Category Theorem for a complete metric space (Theorem 1.6.1), with neighborhoods having compact closures replacing the open balls used there. See Exercise 6.

Definition 3.5.9. If X is locally compact and $\phi : X \to \mathbb{R}$ is a continuous function, say that ϕ *vanishes at infinity* if for every $\epsilon > 0$, $\{x \in X : |\phi(x)| \geq \epsilon\}$ is compact. For any continuous function $\phi : X \to \mathbb{R}$, define the *support* of ϕ, in symbols $\mathrm{spt}\, \phi$, as the set

$$\mathrm{spt}\, \phi = \mathrm{cl}\, \{x \in X : \phi(x) \neq 0\}.$$

A continuous function ϕ on X is said to have *compact support* if $\mathrm{spt}\, \phi$ is a compact set. We denote by $C_0(X)$ the set of all continuous functions on X that vanish at infinity and by $C_c(X)$ the set of continuous functions on X having compact support.

The reader may have noticed that the definition of a function vanishing at infinity does not need the underlying topological space to be locally compact. It follows that if there are many continuous functions that vanish at infinity, then there are many open sets with compact closures; and this is close to having X locally compact. In fact, if there is a nonzero function ϕ on X with compact support, then for every $\epsilon > 0$, $\{x \in X : |\phi(x)| \geq \epsilon\}$ is a compact set with an interior that at least includes the set $\{x \in X : |\phi(x)| > \epsilon\}$. So if we want to assume that there are plenty of continuous functions that vanish at infinity, there must be many open sets with compact closure. So to simplify our lives, we will never discuss $C_0(X)$ unless X is locally compact. The same remarks apply to $C_c(X)$.

Example 3.5.10. Let X be a locally compact space.
(a) The nonzero constant functions on X do not have compact support and do not vanish at infinity unless X is compact.
(b) $C_c(X) \subseteq C_0(X)$, and Theorem 3.5.5 shows the existence of many functions in $C_c(X)$.
(c) A function ϕ belongs to $C_0(\mathbb{R})$ if and only if $\lim_{x \to \pm\infty} \phi(x) = 0$.
(d) If $\phi(x) = \exp(-x^2)$ for x in $\mathbb{R}$, then $\phi \in C_0(\mathbb{R})$, but $\phi \notin C_c(\mathbb{R})$.

Proposition 3.5.11. *Let X be a locally compact space.*
(a) *Both $C_0(X)$ and $C_c(X)$ are subalgebras of $C_b(X)$.*
(b) *$C_0(X)$ is closed in $C_b(X)$.*
(c) *$C_c(X)$ is dense in $C_0(X)$.*

Proof.   (a) The fact that $C_0(X) \subseteq C_b(X)$ is straightforward. In fact, if $\phi \in C_0(X)$, let K be a compact subset of X such that $|\phi(x)| < 1$ when $x \notin K$. Since K is compact, there is a constant M such that $|\phi(x)| \leq M$ whenever $x \in K$. Thus, $|\phi(x)| \leq \max\{M, 1\}$ for all x, and so $\phi \in C_b(X)$. If $\phi_1, \phi_2 \in C_0(X)$ and $\epsilon > 0$, then there are compact sets K_1, K_2 such that for $j = 1, 2$ and $x \in X \backslash K_j$, $|\phi_j(x)| < \epsilon/2$. Therefore, for x in $X \backslash (K_1 \cup K_2)$, $|\phi_1(x) + \phi_2(x)| \leq |\phi_1(x)| + |\phi_2(x)| < \epsilon$. Thus, $\phi_1 + \phi_2 \in C_0(X)$. The proof of the rest of (a) is Exercise 8.

(b) Let $\{\phi_n\}$ be a sequence in $C_0(X)$, and suppose $f \in C_b(X)$ such that $\phi_n \to f$. Let $\epsilon > 0$, and choose N such that $|\phi_n(x) - f(x)| < \epsilon/2$ when $n \geq N$. Let $K = \{x : |\phi_N(x)| \geq \epsilon/2\}$. If $x \in X \backslash K$, then $|f(x)| \leq |f(x) - \phi_N(x)| + |\phi_N(x)| < \epsilon$. Hence $f \in C_0(X)$.

(c) Let $\phi \in C_0(X)$, and assume that $\phi(X) \subseteq [a, b]$. If $\epsilon > 0$, let K be a compact set such that $|\phi(x)| < \epsilon/2$ when $x \notin K$. Let G be an open set with compact closure that contains K (Lemma 3.5.4), and let $\psi : X \to [0, 1]$ be a continuous function with $\psi(x) = 1$ for x in K and $\psi(x) = 0$ for x in $X \backslash G$. So $\phi(x)\psi(x) = 0$ when $x \in X \backslash G$ and, hence, $\phi\psi \in C_c(X)$. Now $|\psi(x)\phi(x) - \phi(x)| = 0$ for x in K and is less than $\epsilon/2$ when $x \notin G$. When $x \in G \backslash K$, $|\psi(x)\phi(x) - \phi(x)| \leq |\psi(x)\phi(x)| + |\phi(x)| \leq 2|\phi(x)| < \epsilon$. Hence, $\rho(\phi, \psi\phi) < \epsilon$, and we have that $C_c(X)$ is dense in $C_0(X)$.   ■

When the locally compact space is metrizable, we get a bonus for the functions that vanish at infinity.

Proposition 3.5.12. *If (X, d) is a locally compact metric space, then every function in $C_0(X)$ is uniformly continuous.*

Proof. Let $\epsilon > 0$, and put $L = \{x : |\phi(x)| \geq \epsilon/2\}$; so L is compact. Since X is locally compact, for every x in L there is an $r_x > 0$ such that $\operatorname{cl} B(x; r_x)$ is compact [Proposition 3.5.3(d)]. Let $x_1, \ldots, x_m \in L$ such that $L \subseteq \bigcup_{j=1}^{m} B(x_j; r_{x_j})$. Choose $\gamma > 0$ such that $\operatorname{dist}(x, L) \leq \gamma$ implies $x \in \bigcup_{j=1}^{m} B(x_j; r_{x_j})$ (Exercise 1.4.7), and set $K = \{x : \operatorname{dist}(x, L) \leq \gamma\}$. Note that K is compact since it is contained in $\bigcup_{j=1}^{m} \operatorname{cl} B(x_j; r_{x_j})$. If we only consider ϕ as a function on K, it is uniformly continuous there; so there exists δ_1 such that $|\phi(x) - \phi(y)| < \epsilon$ when $x, y \in K$ and $d(x, y) < \delta_1$. Choose a positive δ with $\delta < \min\{\delta_1, \gamma\}$. Let $x, y \in X$ such that if $x, y \in L$, then it must be that $x, y \in K$, and so $|\phi(x) - \phi(y)| < \epsilon$. If $x \in L$ but $y \notin L$, then the fact that $d(x, y) < \delta < \gamma$ implies that $x, y \in K$; hence, $|\phi(x) - \phi(y)| < \epsilon$. If neither point belongs to L, then $|\phi(x) - \phi(y)| < |\phi(x)| + |\phi(y)| < \epsilon/2 + \epsilon/2 - \epsilon$. ∎

We now turn our attention to embedding a locally compact space inside a compact one. If the reader has seen § 3.4, that presents one way of doing this since every locally compact space is completely regular. But here we seek a simpler way, which, of course, does not possess all the properties of βX but has other virtues—especially simplicity. The process is related to Example 3.5.3(d). Specifically, if $X = (0, 1]$, then the embedding will be into $[0, 1]$, which, as was pointed out in § 3.4, is far different than βX.

Before giving the definition, we prove a result that justifies it. But first, let us take a moment to see how we add a single point to a given abstract set X. What we want to do is just add it: we are given a set X and take some abstract point—for lack of a better name we call it ∞—and we consider the set $X \cup \{\infty\}$. If you are comfortable with that, go directly to the next proposition and do not worry about the intervening material. If you are bothered by taking the union of these two sets X and $\{\infty\}$ when they are not both subsets of a common set, let us take a moment to allay your concerns.

There are many ways to do this, and all are equivalent. Here is one way. First take two points anywhere—say, the points 0 and 1 in $\mathbb{R}$—and look at $X \times \{0, 1\}$. Now fix a point y in X and consider the set $Z = X \times \{0\} \cup \{(y, 1)\} \subseteq X \times \{0, 1\}$. The set Z contains a copy of X, namely $X \times \{0\}$, and $Z \setminus X$ is a single point, $\{(y, 1)\}$. So $X \times \{0\}$ and X are identified, and we call the point $(y, 1)$ by the name ∞. Thus we have added a single point to X.

My advice is to be practical, abandon this formality, and just be direct. Adopt the casual approach described previously. That is what we do in what follows.

Proposition 3.5.13. *If X is a locally compact space, then there is a compact space X_∞ having the following properties.*

(a) $X \subseteq X_\infty$ and $X_\infty \backslash X$ is a single point denoted by ∞;

(b)   A subset U of X_∞ is open if and only if the following two conditions are satisfied: (i) $U \cap X$ is open in X; (ii) if $\infty \in U$, then $X_\infty \backslash U$ is a compact subset of X;

(c)   If $\phi \in C_0(X)$ and we define $\tilde{\phi} : X_\infty \to \mathbb{R}$ by setting $\tilde{\phi}(x) = \phi(x)$ when $x \in X$ and $\tilde{\phi}(\infty) = 0$, then $\tilde{\phi} \in C(X_\infty)$. Conversely, if $f \in C(X_\infty)$ such that $f(\infty) = 0$ and $\phi = f|X$, then $\phi \in C_0(X)$ and $\tilde{\phi} = f$.

Note that part (b) of the proposition says that the relative topology that X has as a subset of X_∞ is its original topology. Equivalently, X is homeomorphically contained in X_∞.

Proof. Let ∞ be an abstract point, and let $X_\infty = X \cup \{\infty\}$. Let $\mathcal{U}$ be the collection of all subsets U of X_∞ that satisfy the condition stated in (b). We leave it to the reader to verify that $\mathcal{U}$ is a topology on X_∞ (including that it has the Hausdorff property). Now to show that $(X_\infty, \mathcal{U})$ is compact. If $\mathcal{C}$ is an open cover of X_∞, then there is a set U_0 in the cover $\mathcal{C}$ such that $\infty \in U_0$. By definition of the topology $\mathcal{U}$, $K = X_\infty \backslash U_0$ is compact. Thus, there are sets $U_n, \ldots, U_n$ in $\mathcal{C}$ such that $X_\infty \backslash U_0 \subseteq \bigcup_{k=1}^n U_k$. Therefore, $\{U_0, U_1, \ldots, U_n\}$ is the sought-after finite subcover of $\mathcal{C}$, and we have established that X_∞ is compact. This proves parts (a) and (b) of the proposition.

To show that (c) holds is not difficult. If $\phi \in C_0(X)$ and $\tilde{\phi}$ is as in (c), then for any open subset G of $\mathbb{R}$ that does not contain 0 we have that $\tilde{\phi}^{-1}(G) = \phi^{-1}(G)$, an open subset of X; thus, $\tilde{\phi}^{-1}(G) \in \mathcal{U}$. If $r > 0$, then $X_\infty \backslash \tilde{\phi}^{-1}((-r,r)) = \{x \in X : |\phi(x)| \geq r\}$ is compact, so $\tilde{\phi}^{-1}((-r,r)) \in \mathcal{U}$. Thus, $\tilde{\phi}$ is continuous on X_∞. Now assume that $f \in C(X_\infty)$ with $f(\infty) = 0$, and put $\phi = f|X$. It is immediate from the definition of the relative topology on X that $\phi \in C_b(X)$. On the other hand, for any $\epsilon > 0$, the fact that $f(\infty) = 0$ implies that $\{x \in X : |\phi(x)| \geq \epsilon\} = X \backslash f^{-1}((-\epsilon, \epsilon))$, which is compact. Thus, $\phi \in C_0(X)$. The fact that $\tilde{\phi} = f$ is clear. ∎

Definition 3.5.14. If X is a locally compact space, the topological space X_∞ in the preceding proposition is called the *one-point compactification* of X.

See Exercise 10.

Example 3.5.15.   (a) The one-point compactification of $\mathbb{R}$ is homeomorphic to the circle $\{(x,y) \in \mathbb{R}^2 : x^2 + y^2 = 1\}$. It is not $[-\infty, \infty]$ because this requires adding two points.

(b) The one-point compactification of $(0,1]$ is the closed unit interval. The one-point compactification of $(0,1)$ is the circle.

(c) The one-point compactification of the plane is the sphere in $\mathbb{R}^3$: $\{(x,y,z) \in \mathbb{R}^3 : x^2 + y^2 + z^2 = 1\}$.

The one-point compactification has many uses. For example, when we want to prove something about a locally compact space, we can first prove it for compact spaces and then show that it holds for the open subset of a compact space that results from removing a single point. In effect, that is what was going on when we proved the analogues of Urysohn's Lemma (Theorem 3.5.5) and Tietze's Theorem (Theorem 3.5.7) to locally compact spaces. See Exercise 11.

Nevertheless, the only topological property we will investigate about the one-point compactification is the question of when it is metrizable. That is, when is there a metric on X_∞ such that the topology defined by the metric is the topology of X_∞? (See the end of §2.1.) As it turns out, we will see that this question is related to another question that has importance. Several necessary conditions are revealed after a few minutes' thought. First is that when X_∞ is metrizable, then X is metrizable since the topology of X is the relative topology it gets as a subspace of X_∞. Second, if (X, d) is a locally compact metric space and X_∞ is metrizable, then the metric on X will not necessarily be the restriction of the metric from X_∞. For example, as we pointed out in Example 3.5.15(a), the one-point compactification of $\mathbb{R}$ is homeomorphic to the circle. For a compact metric space, the metric must be bounded. That is, if (Z, η) is a compact metric space, then there is a constant M such that $\eta(z, w) \leq M$ for all z, w in Z. So if X_∞ is metrizable, then all we can conclude is that the metric on X is equivalent (Definition 1.3.12) to the metric it inherits from X_∞. Finally, note that if X_∞ is metrizable and η is a metric on X_∞ that defines its topology, then for any $r > 0$, $X \backslash B(\infty; r)$ is a compact subset of X since $B(\infty; r)$ is an open subset of X_∞ that contains ∞. By taking $r = 1/n$, we see that X can be written as the union of a sequence of compact sets. We isolate this property.

Definition 3.5.16. Say that a topological space X is *σ-compact* if it is the union of a sequence of compact sets.

We note that $\mathbb{R}^q$ is σ-compact, as is every open subset of Euclidean space (Exercise 16). If X is locally compact and σ-compact, then we can write X as the union of compact sets K_n such that $K_n \subseteq \operatorname{int} K_{n+1}$ (Exercise 17). The proof of the next theorem is long and complicated, though each step is not so difficult. Just take it slow and understand each step. After going through the proof once, go back over it and try to get the big picture. In other words, first examine all the trees and then step back and look at the forrest. You will be better off for the double effort. In this proof, we will see our first significant use of a partition of unity (Corollary 3.5.11) and perhaps begin to appreciate the power of this tool.

Theorem 3.5.17. *If (X, d) is a locally compact metric space, the following statements are equivalent.*

(a) X_∞ *is metrizable.*

(b) X *is σ-compact.*

(c)   *The metric space $C_0(X)$ is separable.*

Proof. (a) *implies* (b). This was already shown prior to the statement of the theorem.

(b) *implies* (c). We write $X = \bigcup_{n=1}^{\infty} K_n$, where each K_n is compact and $K_n \subseteq \text{int } K_{n+1}$ for all $n \geq 1$ (Exercise 17). Find a decreasing sequence of positive numbers $\{\delta_n\}$ such that $\{x : \text{dist}(x, K_n) < \delta_n\} \subseteq \text{int } K_{n+1}$. (How?) For each n and each $k \geq n$, let $\{B(a_{nk}^j; \delta_k) : 1 \leq j \leq m_{nk}\}$ be open disks with a_{nk}^j in K_n such that $K_n \subseteq \bigcup_{j=1}^{m_{nk}} B(a_{nk}^j; \delta_k)$. Note that for $k \geq n$ this union of open disks is a subset of $\text{int } K_{n+1}$. As in Corollary 3.3.11, for each $n \geq 1$ and $k \geq n$, let $\{\phi_{nk}^j : 1 \leq j \leq m_{nk}\}$ be continuous functions on X with $0 \leq \phi_{nk}^j \leq 1$, $\phi_{nk}^j(x) = 0$ for $x \notin B(a_{nk}^j; \delta_k)$, $\sum_{j=1}^{m_{nk}} \phi_{nk}^j(x) = 1$ when $x \in K_n$, and $\sum_{j=1}^{m_{nk}} \phi_{nk}^j \leq 1$. Let $\mathcal{M}$ be the linear span of the functions $\{\phi_{nk}^j : n \geq 1, k \geq n, \text{ and } 1 \leq j \leq m_{nk}\}$ with coefficients from the rational numbers $\mathbb{Q}$. Note that $\mathcal{M}$ is a countable subset of $C_c(X)$. We will show that $\mathcal{M}$ is dense in $C_c(X)$ and, hence, in $C_0(X)$ (Proposition 3.5.11).

Fix f in $C_c(X)$, and let $\epsilon > 0$; so there is an integer n such that $f(x) = 0$ when $x \notin K_n$. Since f is uniformly continuous, there is a $\delta > 0$ such that $|f(x) - f(y)| < \epsilon/2$ whenever $d(x, y) < \delta$. Pick $k \geq n$ such that $\delta_k < \delta$. For $1 \leq j \leq m_{nk}$ let $q_{nk}^j \in \mathbb{Q}$ such that $|q_{nk}^j - f(a_{nk}^j)| < \epsilon/2$. Hence, $g = \sum_{j=1}^{m_{nk}} q_{nk}^j \phi_{nk}^j \in \mathcal{M}$. Now fix x in K_n. Thus (reader: pay attention here to how the partition of unity is used),

$$|f(x) - g(x)| = \left| \sum_{j=1}^{m_{nk}} [f(x) - q_{nk}^j] \phi_{nk}^j(x) \right|$$

$$\leq \sum_{j=1}^{m_{nk}} \left| f(x) - f(a_{nk}^j) \right| \phi_{nk}^j(x) + \sum_{j=1}^{m_{nk}} \left| f(a_{nk}^j) - q_{nk}^j \right| \phi_{nk}^j(x)$$

$$\leq \sum_{j=1}^{m_{nk}} \left| f(x) - f(a_{nk}^j) \right| \phi_{nk}^j(x) + \epsilon/2.$$

Now when $\phi_{nk}^j(x) \neq 0$, $x \in B(a_{nk}^j; \delta_k)$, and so $\left| f(x) - f(a_{nk}^j) \right| < \epsilon/2$. That is, each term in the last sum is either 0 or, if not, smaller than $\epsilon/2$. Thus, the sum is dominated by

$$\sum_{j=1}^{m_{nk}} (\epsilon/2) \phi_{nk}^j(x) = \epsilon/2.$$

Inserting this inequality into the preceding one yields that $|f(x) - g(x)| < \epsilon$ when $x \in K_n$. Suppose now that $x \in \left[\bigcup_{j=1}^{m_{nk}} B(a_{nk}^j; \delta_k) \right] \setminus K_n$. So $f(x) = 0$. If $x \in B(a_{nk}^j; \delta_k)$, then, by the uniform continuity condition on f, $\left| f(a_{nk}^j) \right| = \left| f(a_{nk}^j) - f(x) \right| < \epsilon/2$. Therefore,

$$|f(x) - g(x)| = |g(x)|$$

$$\leq \sum_{j=1}^{m_{nk}} \left| q_{nk}^j \right| \phi_{nk}^j(x)$$

$$\leq \sum_{j=1}^{m_{nk}} \left[\left| q_{nk}^j - f(a_{nk}^j) \right| + \left| f(a_{nk}^j) \right| \right] \phi_{nk}^j(x)$$

$$< (\epsilon/2 + \epsilon/2) \sum_{j=1}^{m_{nk}} \phi_{nk}^j(x).$$

$$\leq \epsilon$$

Finally, if $x \notin \bigcup_{j=1}^{m_{nk}} B(a_{nk}^j; \delta_k)$, then $f(x) = 0 = g(x)$. Therefore, if ρ denotes the metric on $C_0(X)$, then we have that $\rho(f, g) < \epsilon$, and so $\mathcal{M}$ is dense in $C_0(X)$.

(c) *implies* (a). Here is an outline of the proof, which will be followed by the details. First note that when $\phi \in C_0(X)$, $\eta_\phi(x, y) = |\phi(x) - \phi(y)|$ is symmetric ($\eta_\phi(x, y) = \eta_\phi(y, x)$) and satisfies the triangle inequality. η_ϕ is called a semimetric. Extending ϕ to X_∞ by setting $\phi(\infty) = 0$ enables us to see that η_ϕ defines a semimetric on X_∞. Now using the fact that $C_0(X)$ is separable we can use a countable dense subset of $B = \{\phi \in C_0(X) : |\phi(x)| \leq 1 \text{ for all } x\}$ to generate a sequence of such semimetrics and sum them up to get a true metric on X_∞. We will then show that this metric defines the topology on X_∞. Now for the details.

Let $\{\phi_n\}$ be a countable dense sequence in the subset B of $C_0(X)$ defined in the last paragraph, and define $\eta(x, y) = \sum_{n=1}^{\infty} 2^{-n}|\phi_n(x) - \phi_n(y)|$ for all x, y in X_∞. Note that for x in X, $\eta(x, \infty) = \sum_{n=1}^{\infty} 2^{-n}|\phi_n(x)|$ since $\phi_n(\infty) = 0$ for all n. Clearly, $\eta(x, y) = \eta(y, x)$. If $\eta(x, y) = 0$, then $\phi_n(x) = \phi_n(y)$ for all $n \geq 1$. Assume $x \neq y$; provided $x \neq \infty$, there is a function ϕ in $C_c(X)$ such that $\phi(x) = 1$ and $\phi(y) = 0$. (Why?) Let $n \geq 1$ such that $\rho(\phi, \phi_n) < \frac{1}{2}$. It follows that $|\phi_n(x)| > \frac{1}{2}$ and $|\phi_n(y)| < \frac{1}{2}$. Thus, it cannot be that $\eta(x, y) = 0$. We leave it to the reader to verify that the triangle inequality holds for η so that it is a metric on X.

It remains to prove that this metric η on X_∞ defines its topology. To accomplish this, we will show that the inclusion map $(X, d) \to (X_\infty, \eta)$ is a homeomorphism onto its image and that $\{x : \eta(x, \infty) \geq \epsilon\}$ is compact for every $\epsilon > 0$. (Are you clear that this will do the job?)

Claim. A sequence $\{x_n\}$ converges to x in (X_∞, η) if and only if $\phi_k(x_n) \to \phi_k(x)$ for all $k \geq 1$.

If $x_n \to x$ in (X_∞, η), then the fact that $2^{-k}|\phi_k(x_n) - \phi_k(x)| \leq \eta(x_n, x)$ implies $\phi_k(x_n) \to \phi_k(x)$ for all $k \geq 1$. Now assume that $\phi_k(x_n) \to \phi_k(x)$ for all $k \geq 1$, and let $\epsilon > 0$. Choose m such that $\sum_{k=m}^{\infty} \frac{1}{2^k} < \epsilon/2$, and choose N such that for $n \geq N$ and $1 \leq k \leq m$, $|\phi_k(x_n) - \phi_k(x)| < \epsilon/2m$. Thus, for

$n \geq N$,

$$\eta(x_n, x) < \epsilon/2 + \sum_{k=1}^{m} \frac{1}{2^k} \frac{\epsilon}{2m} < \epsilon,$$

proving the claim.

Claim. The inclusion map $(X, d) \to (X_\infty, \eta)$ is a homeomorphism.

The first claim proves that if $x_n \to x$ in X, then $\eta(x_n, x) \to 0$; that is, the inclusion map $(X, d) \to (X_\infty, \eta)$ is continuous. For the converse, suppose that $\eta(x_n, x) \to 0$, where x and x_n belong to X for all $n \geq 1$. Suppose that $\{d(x_n, x)\}$ does not converge to 0; then there is an $\epsilon > 0$ and a subsequence $\{x_{n_j}\}$ such that $d(x_{n_j}, x) \geq \epsilon$ for all n_j. We can assume that ϵ is small, say $\epsilon < \frac{1}{2}$. Using Theorem 3.5.5, there is a function ϕ in $C_c(X)$ such that $0 \leq \phi \leq 1$, $\phi(x) = 1$, and $\phi(y) = 0$ when $d(y, x) \geq \epsilon/2$. Since $\{\phi_k\}$ is dense in B, there is a ϕ_k such that $\rho(\phi_k, \phi) < \epsilon/2$. Hence, $|\phi_k(x_{n_j})| = |\phi_k(x_{n_j}) - \phi(x_{n_j})| < \epsilon/2$. On the other hand, $1 = \phi(x) \leq |\phi(x) - \phi_k(x)| + |\phi_k(x)| < \epsilon/2 + |\phi_k(x)|$, and so $|\phi_k(x)| > 1 - \epsilon/2 > \epsilon/2$. This contradicts the assumption that $\phi_k(x_{n_j}) \to \phi_k(x)$, which, by the first claim, contradicts the assumption that $\eta(x_n, x) \to 0$. This establishes the second claim.

Claim. $\{x : \eta(x, \infty) \geq \epsilon\}$ is compact for every $\epsilon > 0$.

Let $\epsilon > 0$ and put $K = \{x \in X : \eta(x, \infty) \geq \epsilon\}$. Suppose K is not compact. For each $n \geq 1$, put $K_n = \{x \in X : |\phi_k(x)| \geq \frac{1}{n}$ for $1 \leq k \leq n\}$. Since K is closed and each K_n is compact, it cannot be that $K \subseteq K_n$. Therefore, there is a point x_n in K such that $x_n \notin K_n$. That is, $|\phi_k(x_n)| < \frac{1}{n}$ for $1 \leq k \leq n$. This says that for every $k \geq 1$, $\lim_n \phi_k(x_n) = 0$. According to the first claim, this implies that $x_n \to \infty$ or that $\eta(x_n, \infty) \to 0$. Since each $x_n \in K$, this is a contradiction. Therefore, K must be compact.

This completes the proof.   ∎

Corollary 3.5.18. *If X is a compact topological space, then $C(X)$ is separable if and only if X is a compact metric space.*

Proof. The fact that X is a compact space implies $X = X_\infty$. Thus, the corollary is immediate from the theorem.   ∎

Exercises

(1) Show that $X = \{(0,0)\} \cup \{(x, y) : x > 0, y \in \mathbb{R}\}$ with the relative topology from $\mathbb{R}^2$ is not locally compact. So an arbitrary subset of a locally compact space is not necessarily locally compact.

(2) (a) Show that a dense subset of a locally compact space is locally compact if and only if it is open. (This gives another proof that $\mathbb{Q}$ is not locally compact.) (b) Show that a subset E of a locally compact space X is locally compact if and only if $E = A \backslash B$, where both A and B are closed subsets of X.

(3) Finish the proof of Proposition 3.5.3(b).

(4) Finish the proof of Proposition 3.5.3(c).

(5) Give the details of the proof of Proposition 3.5.3(d).

(6) Prove the Baire Category Theorem for locally compact spaces.

(7) If X is a locally compact space, K is a compact subset, G is an open set with $K \subseteq G$, and $f \in C_b(X)$ with $a \leq f(x) \leq b$ for all x in X, then there is a continuous function $F : X \to [a, b]$ such that $F(x) = f(x)$ for all x in K and $F(x) = 0$ when $x \notin G$.

(8) Complete the proof of Proposition 3.5.11(a).

(9) If X is a locally compact space and $f, g \in C_0(X)$, show that $f \vee g$ and $f \wedge g \in C_0(X)$. Similarly, if $f, g \in C_c(X)$, show that $f \vee g$ and $f \wedge g \in C_c(X)$.

(10) Phrase and prove a proposition that shows that the one-point compactification of a locally compact space is unique up to a homeomorphism.

(11) If X is a locally compact space, use the one-point compactification of X to give proofs of Theorems 3.5.5 and 3.5.7 as a consequence of Urysohn's Lemma and the Tietze Extension Theorem.

(12) What is the one-point compactification of $\mathbb{R}^q$?

(13) Describe the one-point compactification of $\mathbb{N}$ by finding a subset of $\mathbb{R}$ that is homeomorphic to it.

(14) Let I be any nonempty set, and for each i in I let X_i be a copy of $\mathbb{R}$ with the metric $d_i(x, y) = |x - y|$. Let X be the disjoint union of the sets X_i. (That is a verbal description that can be used in any circumstance, but if you want precision, you can say $X = \mathbb{R} \times I$, the cartesian product, where I has the discrete topology.) Define a metric on X by letting d agree with d_i on each X_i; and when $x \in X_i$, $y \in X_j$, where $i \neq j$, then define $d(x, y) = 1$. (a) Show that d is indeed a metric on X. (b) Show that $\{X_i : i \in I\}$ is the collection of components of X and each of these components is an open subset of X. (c) Show that (X, d) is separable if and only if I is a countable set.

(15) Consider the metric space (X, d) defined in Exercise 14, and assume that I is not countable. (a) Show that (X, d) is locally compact. (b) Show that (X, d) is not σ-compact and, hence, X_∞ is nonmetrizable. (c) Find an infinite subset A of X such that ∞ is in the closure of A in X_∞ but no sequence of points from A converges to ∞ in the one-point compactification.

(16) If X is a locally compact, σ-compact topological space, show that every open subset and every closed subset with the relative topology is also σ-compact. [See Proposition 3.5.3(b).]

(17) Assume that X is locally compact and σ-compact. (a) Show that we can write X as the union of compact sets K_n such that $K_n \subseteq \operatorname{int} K_{n+1}$. (b) Show that there is a sequence of functions $\{\phi_n\}$ in $C_c(X)$ such that for every ϕ in $C_0(X)$, $\phi_n \phi \to \phi$ in the metric of $C_0(X)$.

(18) If X is locally compact, show that X is σ-compact if and only if there is a function ϕ in $C_0(X)$ such that $X = \{x : \phi(x) \neq 0\}$.

3.6. Ordinal Numbers*

The purpose of this short section is to introduce two topological spaces that are very useful for constructing examples. The building of these spaces relies on the concept of an ordinal number. For a precise exposition of this topic the reader is referred to [4] and to the appendix of [6]. There are many other sources. The approach here is to outline the definition of ordinal numbers and then state the basic properties needed for the progress of this section. We omit the proofs of those properties that would take us too far from our main objective. The reader can take these unproven statements as axioms or consult the references for the proofs.

The definition of ordinal numbers is similar to that of cardinal numbers. We consider all *well-ordered sets*, that is, a linearly ordered set (§ A.4) $(S, \leq)$ such that if E is a nonempty subset of S, then E contains a least element. Say that two such sets are equivalent if there is an order isomorphism between them; that is, there is a bijection between the two sets that preserves the order. An *ordinal number* is an equivalence class of such sets. (There are some logical problems here in considering the collection of *all* such sets. Nevertheless, ignoring such problems will not cause any difficulty for us. If you want to delve into this, consult a friendly logician.) If two well-ordered sets are equivalent, then we may say that they have the same *order type*.

It is not difficult to give examples of such sets. Every finite set is easily made into a well-ordered set, and no matter how we do this, two finite, well-ordered sets are equivalent if and only if they have the same number of elements. The set $\mathbb{N}$ is a well-ordered set, but $\mathbb{Z}$ and $\mathbb{R}$ with their natural ordering are not. (Why?) Other examples of well-ordered sets are $\{1 - \frac{1}{n} : n \in \mathbb{N}\} \cup \{1\}$, $\{2 - \frac{1}{n} : n \in \mathbb{N}\} \cup \{2\}$, $\{1 - \frac{1}{n} : n \in \mathbb{N}\} \cup \{2 - \frac{1}{n} : n \in \mathbb{N}\} \cup \{1, 2\}$, and $\{m - \frac{1}{n} : m, n \in \mathbb{N}\} \cup \mathbb{N}$, with the order they have as subsets of $\mathbb{R}$.

For the rest of this section we only consider well-ordered sets $(S, \leq)$.

We might pause to look at these examples and observe that nonempty subsets of a well-ordered set do not always contain a largest element even though they contain a smallest one. Nevertheless, we want to define $\sup E$ for a nonempty subset E of S. When $\{y \in S : y \geq x \text{ for all } x \in E\} = \emptyset$, this is not possible, as a moment's reflection will show. When this is not the case, however, we define $\sup E$ as the smallest element in the set $\{y \in S : y \geq x \text{ for all } x \in E\}$. What we just observed prior to this definition is that when $\sup E$ exists, it may not belong to E. We will denote the always extant smallest element of E as $\inf E$ or $\min E$.

We note that the sets $\{1 - \frac{1}{n} : n \in \mathbb{N}\} \cup \{1\}$ and $\{2 - \frac{1}{n} : n \in \mathbb{N}\} \cup \{2\}$ have the same order type, while $\{1 - \frac{1}{n} : n \in \mathbb{N}\} \cup \{1\}$ and $\{1 - \frac{1}{n} : n \in \mathbb{N}\} \cup \{2 - \frac{1}{n} : n \in \mathbb{N}\} \cup \{1, 2\}$ do not. While the positive of these two statements is easy to see and the negative one is intuitively clear, one way to prove the negative statement is as follows.

We give a well-ordered set $(S, \leq)$ the order topology (Exercise 2.2.5). In this situation the order topology on S is Hausdorff. In fact, if $x < y$ in S, then $U = \{z \in S : z < y\}$ and $V = \{z \in S : x < z\}$ are disjoint open sets, $x \in U$ and $y \in V$. Now note that an order isomorphism between two well-ordered sets is a homeomorphism when they are given their order topology. Since the set $\{1 - \frac{1}{n} : n \in \mathbb{N}\}$ has one limit point while the set $\{1 - \frac{1}{n} : n \in \mathbb{N}\} \cup \{2 - \frac{1}{n} : n \in \mathbb{N}\}$ has two, they cannot have the same order type.

When we have a well-ordered set $(S, \leq)$, we will often refer to its elements as ordinals. In fact, if $x \in S$, then the set $\{y \in S : y \leq x\}$ is a well-ordered set and we can take x as its order type. The last paragraph illustrates that there are two distinct types of ordinals in $(S, \leq)$: *limit ordinals* and *nonlimit* or *discrete ordinals*. Say that x is a limit ordinal in $(S, \leq)$ precisely when it is a limit point in the order topology. Using the order this means x is a limit ordinal if and only if for any y in S with $y < x$ there is a z in S with $y < z < x$. Say that x is a discrete ordinal when it is not a limit point in the order topology or, equivalently, when it is an isolated point of the topological space. Thus, x is a discrete ordinal if and only if the singleton $\{x\}$ is an open set. Using the order, this means that x is a discrete ordinal if and only if there is a $y < x$ such that $(y, x) = \emptyset$; in other words, x has an immediate predecessor in the order. We denote the immediate predecessor by $x - 1$.

Some additional notation is useful. In $(S, \leq)$, if $a \leq b$, we will give the sets $(a, b), [a, b], (a, b]$, and $[a, b)$ their obvious meaning. Also, if $x \in S$ and $\{y \in S : x < y\} \neq \emptyset$, then this set has a least element and we denote it by $x + 1$. That is, $x + 1$ is the first element in S bigger than x.

It is also true that the set of ordinal numbers is itself a well-ordered set—a fact that needs a proof that will not be given here. If we consider all the ordinal numbers that are uncountable, then it follows that there is a least such ordinal number and we denote it by Ω. (Giving an example of an uncountable ordinal is difficult. It can be shown that there is a well ordering of $\mathbb{R}$, but that is very difficult. Another example is the set of all countable ordinals.) The number Ω is called the *first uncountable ordinal.*

The spaces $[0, \Omega]$ of all ordinal numbers that are less than or equal to Ω and $[0, \Omega)$ of all countable ordinals, as topological spaces, are the object of study in this section. We record some fundamental order properties of $[0, \Omega]$ and $[0, \Omega)$.

Proposition 3.6.1. (a) *The spaces $[0, \Omega]$ and $[0, \Omega)$ are uncountable, well-ordered spaces such that if $x \in [0, \Omega)$, then $\{y \in [0, \Omega] : y \leq x\} = \{y \in [0, \Omega) : y \leq x\}$ is a countable set.*

(b)   *If $\emptyset \neq E \subseteq [0, \Omega]$, then $\sup E \in [0, \Omega]$.*

(c)   *If E is a countable subset of $[0, \Omega)$, then $\sup E < \Omega$.*

The spaces $[0, \Omega]$ and $[0, \Omega)$ have many interesting topological properties. We begin with some of the most basic.

Proposition 3.6.2. (a)   *The collection of sets $\{(a,b] : a, b \in [0, \Omega]$ and $a < b\}$ is a base for the topology of $[0, \Omega]$.*

(b)   *The collection of sets $\{(a,b] : a, b \in [0, \Omega)$ and $a < b\}$ is a base for the topology of $[0, \Omega)$.*

(c)   *If $\{x_n\}$ is an increasing sequence in $[0, \Omega]$ and $x = \sup\{x_n\}$, then $x_n \to x$ in the order topology.*

(d)   *A subset F of $[0, \Omega)$ is closed if and only if it is sequentially closed.*

Proof. The proofs of (a) and (b) are similar, and we only prove (b). Note that when $a < b$, then $(a,b] = (a, b+1)$, so each set in $\mathcal{B} = \{(a,b] : a, b \in [0, \Omega)$ and $a < b\}$ is open in $[0, \Omega)$. Now assume G is an open set in $[0, \Omega)$ and $x \in G$. Since the order intervals $\{(a,b) : a, b \in X$ and $a < b\}$ form a base for the topology (Exercise 2.2.5), there is one such interval such that $x \in (a, b) \subseteq G$. But then $x \in (a, x] \subseteq G$.

(c) If G is a neighborhood of x, then by (a) there is a point a with $a < x$ such that $(a, x] \subseteq G$. From the definition of supremum there is an integer N such that $x_n > a$ when $n \geq N$. Thus $x_n \in G$ for all $n \geq N$, proving the statement.

(d) Since one implication is obvious, it is only required to show that F is closed when it is sequentially closed. Let $x \in \operatorname{cl} F$. If x is a discrete ordinal, then $\{x\}$ is open, and it follows that $x \in F$. Assume that x is a limit ordinal and G is a neighborhood of x. By (b), there is a point $a < x$ such that $(a, x] \subseteq G$. By Proposition 3.6.1(a), (a, x) is a countable set; thus, we can find a sequence $\{x_n\}$ in (a, x) such that $x_n < x_{n+1}$ and $x = \sup x_n$. (Why?) By (c), $x_n \to x$. Thus, $x \in F$.   ∎

Theorem 3.6.3. *The following conditions hold.*

(a) $[0, \Omega]$ *is a compact topological space.*

(b)   *A subset of $[0, \Omega)$ is compact if and only if it is closed and order bounded.*

(c) $[0, \Omega)$ *is locally compact but not compact.*

(d) $[0, \Omega)$ *is a normal space.*

Proof. We begin this proof by establishing the following claim.

Claim. If $a < b < \Omega$, then $[a, b]$ is compact.

From the properties of the ordinal numbers we know that $[a, b]$ is a countable set, so any open cover of $[a, b]$ has a countable subcover. Thus, we need only show that a countable open cover $\mathcal{G} = \{G_n : n \in \mathbb{N}\}$ of $[a, b]$ has a finite subcover. Suppose $\mathcal{G}$ has no finite subcover; then for each n in $\mathbb{N}$ we have that $F_n = [a, b] \setminus \bigcup_{k=1}^{n} G_k \neq \emptyset$. If $x_n = \inf F_n$, then $x_n \in F_n$. Also, $F_{n+1} \subseteq F_n$. (Why?) So for $n \geq N$, $x_n \in F_N$ and $x_n \geq x_N$. If $x = \sup x_n$, then it follows that $x \in [a, b]$ and $x \in F_n$ for all n in $\mathbb{N}$. That is, $x \in [a, b] \cap \bigcap_{n=1}^{\infty} F_n = [a, b] \setminus \bigcup_{n=1}^{\infty} G_n$, contradicting the fact that $\mathcal{G}$ is a cover of $[a, b]$.

(a) Using Proposition 2.4.5 we need only show that every cover by sets from the base of the topology has a finite subcover. If $\mathcal{G}$ is an open cover of $[0, \Omega]$ by basic open sets of the form $(a, b]$, fix $(a_0, \Omega]$ in $\mathcal{G}$. Now $[0, \Omega] \backslash (a_0, \Omega] = [0, a_0]$, and the claim says that this latter set is compact. Thus, we can find a finite number of sets in $\mathcal{G}$ that cover $[0, a_0]$, and, together with $(a_0, \Omega]$, this provides the sought-after finite subcover of $\mathcal{G}$. Thus, $[0, \Omega]$ is compact.

(b) If K is closed and bounded in $[0, \Omega)$, then there are $a < b < \Omega$ such that $K \subseteq [a, b]$. Using the claim we have that K is a closed subset of a compact set.

(c) If $x \in [0, \Omega)$, then each set $(a, x]$ is a neighborhood of x, which, by the claim, has compact closure.

(d) Suppose A and B are two disjoint closed sets in $[0, \Omega)$. For each a in A put $b_a = \sup\{b \in B : b < a\}$. Note that $(b_a, a]$ is an open neighborhood of a that is disjoint from B. Thus, if $U = \bigcup\{(b_a, a] : a \in A\}$, then U is an open set, $A \subseteq U$, and $U \cap B = \emptyset$. Similarly for each b in B put $a_b = \sup\{a \in A : a < b\}$, so that $a_b \in A$, and if $V = \bigcup\{(a_b, b] : b \in B\}$, then V is an open set that contains B and is disjoint from A. Now to show that $U \cap V = \emptyset$. In fact, if this is not the case, then there is an a in A and a b in B such that $(b_a, a] \cap (a_b, b] \neq \emptyset$. Let $c \in (b_a, a] \cap (a_b, b]$; so $b_a < c \leq a$ and $a_b < c \leq b$. Now either $b < a$ or $a < b$. If $b < a$, then, by the definition of b_a, we have that $b \leq b_a < c \leq a$. Since we also have that $a_b < c \leq b$, we have a contradiction. Similarly, the assumption that $a < b$ leads to a contradiction. Thus, $U \cap V = \emptyset$.

∎

Now for a truly surprising fact about $[0, \Omega)$.

Proposition 3.6.4. *If $f : [0, \Omega) \to \mathbb{R}$ is a continuous function, then there is a point a in $[0, \Omega)$ such that f is constant on $\{x \in [0, \Omega) : x \geq a\}$.*

Proof. We assume the proposition is false and establish the following claim.

Claim. For every $n \geq 1$ there is an a_n in $[0, \Omega)$ such that whenever $x > a_n$, we have that $|f(x) - f(a_n)| < \frac{1}{n}$.

If this claim is false, then it follows that there is an $n_0 \geq 1$ such that for every z in $[0, \Omega)$ there exists an $x > z$ with $|f(x) - f(z)| \geq 1/n_0$. We now define a sequence $x_0, x_1, \ldots$ inductively. Let $x_0 = 0$. If we have x_n for some $n \geq 0$, let $x_{n+1} = \inf\{x \in (x_n, \Omega) : |f(x) - f(x_n)| \geq 1/n_0\}$, where we know this set is nonempty because we assumed the claim was false. Note that $x_n < x_{n+1}$ and, because f is continuous, $|f(x_{n+1}) - f(x_n)| \geq 1/n_0$. It follows that $c = \sup_n x_n < \Omega$ [Proposition 3.6.1(c)] and $x_n \to c$. Again using the continuity of f, $f(x_n) \to f(c)$, and so $\{f(x_n)\}$ is a Cauchy sequence. But we have that $|f(x_{n+1}) - f(x_n)| \geq 1/n_0$, contradicting the Cauchy condition. This establishes the claim.

If $\{a_n\}$ is the sequence whose existence is given by the claim, then put $a = \sup_n a_n$; again, $a < \Omega$ and $a_n \to a$. But for any x in $[a, \Omega)$ we have that $x > a_n$, and so $|f(x) - f(a_n)| < \frac{1}{n}$. Since $f(a_n) \to f(a)$, we have that $f(x) - f(a)| = 0$. That is, $f(x) = f(a)$ for all $x \geq a$.   ∎

Corollary 3.6.5. *If $f : [0, \Omega) \to \mathbb{R}$ is a continuous function, then f is bounded.*

Corollary 3.6.6. *The Stone–Čech compactification of $[0, \Omega)$ is the space $[0, \Omega]$, which is also its one-point compactification.*

Proof. By the proposition every f in $C([0, \Omega)) = C_b([0, \omega))$ is eventually constant. Thus letting $\tilde{f}(\Omega)$ be this constant implies that this gives a continuous extension of $\tilde{f}$ to $[0, \Omega]$. Thus, $\beta[0, \Omega) = [0, \Omega]$ by the uniqueness part of Theorem 3.4.1.   ∎

3.7. Paracompactness

The definition of paracompactness requires two preliminary concepts.

Definition 3.7.1. If X is a topological space and $\mathcal{S}$ is a collection of subsets of X, then $\mathcal{S}$ is *locally finite* if for each x in X there is a neighborhood U of x such that U meets only a finite number of sets in $\mathcal{S}$. If $\mathcal{S}$ and $\mathcal{D}$ are two collections of subsets of X, then $\mathcal{D}$ is said to be a *refinement* of $\mathcal{S}$ if each D in $\mathcal{D}$ is contained in some set S that belongs to $\mathcal{S}$.

Let us underline a few points about this last definition. First, the sets in $\mathcal{S}$ and $\mathcal{D}$ are not assumed open or closed and neither is assumed to be a cover of X. Most of the time when we use the concepts of locally finite and refinement we will be discussing open covers, but there will be times, especially as we develop this topic, when we want extra leeway.

Note that every finite collection of sets is locally finite and a subcollection of $\mathcal{S}$ is a refinement of $\mathcal{S}$. Also, if $\mathcal{D}$ is a refinement of $\mathcal{S}$ and $\mathcal{E}$ is a refinement of $\mathcal{D}$, then $\mathcal{E}$ is a refinement of $\mathcal{S}$.

Definition 3.7.2. A topological space X is *paracompact* if for every open cover $\mathcal{C}$ of X there is an open cover $\mathcal{D}$ of X that is locally finite and a refinement of $\mathcal{C}$.

Example 3.7.3. (a) Every compact space is paracompact.

(b) If X can be written as the pairwise disjoint union of open subsets $\{G_i : i \in I\}$ such that each G_i with its subspace topology is paracompact, then X is paracompact. In fact, if $\mathcal{C}$ is an open cover of X and $\mathcal{D}_i = \{C \cap G_i : C \in \mathcal{C}, i \in I\}$, then $\mathcal{D}_i$ is a cover of G_i and a refinement of $\mathcal{C}$. Since G_i is paracompact, there is an open cover $\mathcal{E}_i$ of G_i that is a locally finite refinement of $\mathcal{D}_i$. If $\mathcal{E} = \bigcup_i \mathcal{E}_i$, then it can be verified (do it) that $\mathcal{E}$ is a cover of X that is a locally finite refinement of $\mathcal{D} = \bigcup_i \mathcal{D}_i$ and, hence, of the original cover $\mathcal{C}$.

We will see many more examples as we progress. You can interpret paracompactness as a generalization of compactness. This will become amplified as we progress. Far less apparent is that paracompactness is a generalization of metrizability. In fact, we have some work to do before this becomes even a meaningful comment, let alone a verifiable one.

Proposition 3.7.4. *If X is paracompact and F is a closed subset of X, then F is paracompact.*

Proof. If $\mathcal{D}$ is an open cover of F, then $\mathcal{C} = \mathcal{D} \cup \{X \backslash F\}$ is an open cover of X. Thus, there is a locally finite open cover of X, $\mathcal{C}_1$, that is a refinement of $\mathcal{C}$. If $\mathcal{D}_1 = \{U \in \mathcal{C}_1 : U \cap F \neq \emptyset\}$, then it can be verified that $\mathcal{D}_1$ is a locally finite refinement of $\mathcal{D}$ that covers F. ∎

An open subset of a paracompact space is not necessarily paracompact. In fact, the space $[0, \Omega]$ (§ 3.6) is compact, $[0, \Omega)$ is an open subset, but $[0, \Omega)$ is not paracompact (Example 3.7.13).

Proposition 3.7.5. *A paracompact space is normal.*

Proof. Assume X is a paracompact topological space. This proof proceeds in stages. We first show that X is regular and then use this to show it is normal.

To show that X is regular, we will use Proposition 3.2.2 and show that if F is a closed subset of X and $c \notin F$, then there is an open set V such that $F \subseteq V$ and $c \notin \operatorname{cl} V$. To see this, use the fact that X is Hausdorff to obtain that for every point x in F there is a neighborhood G_x of x such that $c \notin \operatorname{cl} G_x$ (Exercise 2.1.7). Thus, $\{G_x : x \in F\}$ is an open cover of F. By the preceding proposition, there is an open cover $\mathcal{U}$ of F that is a locally finite refinement of $\{G_x : x \in F\}$. Put $V = \bigcup \{U : U \in \mathcal{U}\}$. Note that V is an open set that contains F. Exercise 1 implies that since $\mathcal{U}$ is locally finite, $\operatorname{cl} V = \bigcup \{\operatorname{cl} U : U \in \mathcal{U}\}$, something that is far from true for arbitrary unions. By the refinement property, for each U in $\mathcal{U}$ there is an x in F such that $U \subseteq G_x$. Since $c \notin \operatorname{cl} G_x$, $c \notin \operatorname{cl} U$; hence we have that $c \notin \operatorname{cl} V$.

We now prove that X is normal. By Proposition 3.3.2, it suffices to show that if A and B are disjoint closed sets, then there is an open set V such that $B \subseteq V$ and $A \cap \operatorname{cl} V = \emptyset$. For each b in B the regularity of X implies there is an open set O_b such that $b \in O_b \subseteq \operatorname{cl} O_b \subseteq X \backslash A$. Let $\mathcal{U}$ be a locally finite refinement of the open cover $\{O_b : b \in B\}$ of B, and put $V = \bigcup \{U : U \in \mathcal{U}\}$; clearly, $B \subseteq V$. As in the preceding paragraph, it follows that $\operatorname{cl} V = \bigcup \{\operatorname{cl} U : U \in \mathcal{U}\}$; and, since $A \cap \operatorname{cl} O_b = \emptyset$ for every b in B, it follows that $A \cap \operatorname{cl} U = \emptyset$ for every U in $\mathcal{U}$ and so $A \cap \operatorname{cl} V = \emptyset$. ∎

See Example 3.7.13 below for a normal space that is not paracompact. The next result can be found in [7], and some extensions appear in [8].

Theorem 3.7.6 (Michael's[5] Theorem). *If X is a regular topological space such that every open cover of X has a refinement cover $\mathcal{A}$ that can be written as $\mathcal{A} = \bigcup_{n=1}^{\infty} \mathcal{A}_n$, where each $\mathcal{A}_n$ is a locally finite collection, then X is paracompact.*

The proof requires a few lemmas, but before we start that process, a few remarks about the condition and the nature of the result would be helpful. First, note that, even though we started with an open cover of X, the collections $\mathcal{A}_n$ are not required to be covers, nor are the sets in them required to be open; $\mathcal{A}$, however, is required to be a cover. Second, since the refinement $\mathcal{A}$ is not required to be locally finite but only the countable union of locally finite collections, this builds even more flexibility into the condition. It is this lack of restrictions that gives the theorem merit. This result does not give insight into the structure of paracompact spaces but rather provides us a tool for showing that a space is paracompact. Finally, the reader may have observed that the condition here is far weaker than the stated definition, so that we could have stated this as a necessary and sufficient condition for paracompactness. Later we will use this theorem to give several important examples of classes of paracompact spaces.

Here is the first lemma, whose purpose here is to facilitate the proof of the second lemma.

Lemma 3.7.7. *If X is a topological space such that every open cover of X has a locally finite closed cover that is a refinement, then X is paracompact.*

Proof. Let $\mathcal{G}$ be an open cover of X. By hypothesis, there is a closed cover refinement $\mathcal{F} = \{F_i : i \in I\}$ of $\mathcal{G}$ that is locally finite. By the definition of locally finite, for each x in X there is a neighborhood W_x of x such that $W_x \cap F_i \neq \emptyset$ for only a finite number of the closed sets F_i in $\mathcal{F}$. Note that $\mathcal{W} = \{W_x : x \in X\}$ is an open cover of X. Once again, apply the hypothesis to obtain a closed cover $\mathcal{C}$ that is locally finite and a refinement of $\mathcal{W}$. Now for each F_i define

$$V_i = X \setminus \bigcup \{C : \text{ such that } C \in \mathcal{C} \text{ and } C \cap F_i = \emptyset\}.$$

By Exercise 1, V_i is open. Moreover, $F_i \subseteq V_i$ (Verify!), so it follows that $\mathcal{V} = \{V_i : i \in I\}$ is an open cover of X.

Claim. $\mathcal{V}$ is a locally finite open cover.

To see this, examine what it means to have a C in $\mathcal{C}$ such that $V_i \cap C \neq \emptyset$. In fact, from the definition of V_i this can only happen when $F_i \cap C \neq \emptyset$. On the other hand, each C in $\mathcal{C}$ is contained in some W_x; since each W_x intersects only a finite number of the sets in $\mathcal{F}$, C can intersect only a finite number

---

[5]Ernest Michael was an American mathematician who was born in Zurich in 1925. He spent most of his career at the University of Washington, where he published over 100 papers, focusing his research efforts on an examination of paracompactness. Many of the deepest results on this topic are due to him. Besides this theorem, he is well known as the author of the Michael Selection Theorem. He produced five doctoral students. He died in Seattle in 2013.

of F_i. Because $\mathcal{C}$ is a locally finite cover of X, this implies $\mathcal{V}$ is locally finite. Now we need to adjust the sets in $\mathcal{V}$ to obtain a locally finite open refinement of the original cover $\mathcal{G}$; this is not hard. For each i in I consider V_i and the corresponding F_i in the closed locally finite cover. Recall that $F_i \subseteq V_i$. Since $\mathcal{F}$ is a refinement of $\mathcal{G}$, there is a set G_i in $\mathcal{G}$ with $F_i \subseteq G_i$. Look at the collection of open sets $\mathcal{U} = \{V_i \cap G_i : i \in I\}$. This is an open cover since each $F_i \subseteq V_i \cap G_i$, and it is locally finite by the claim. Clearly $\mathcal{U}$, from the way it was defined, is a refinement of $\mathcal{G}$, so this completes the proof. ∎

Lemma 3.7.8. *If X is a regular topological space such that for every open cover $\mathcal{G}$ of X there is a cover $\mathcal{C}$ that is a locally finite refinement of $\mathcal{G}$, then X is paracompact, even though it is not required that $\mathcal{C}$ contain either open or closed sets.*

Proof. Let $\mathcal{G}$ be an open cover of X. By the previous lemma we need only show that there is a refinement of $\mathcal{G}$ by a closed locally finite cover. For x in X, let $G_x \in \mathcal{G}$ such that $x \in G_x$. Since X is regular, there is a neighborhood V_x of x with $\operatorname{cl} V_x \subseteq G_x$. Thus, $\mathcal{V} = \{V_x : x \in X\}$ is an open cover that refines $\mathcal{G}$. $\mathcal{V}$ itself may not be locally finite, but, according to the hypothesis, there is a cover $\mathcal{C}$ that is a locally finite refinement of $\mathcal{V}$. Now Exercise 2 implies that $\mathcal{F} = \{\operatorname{cl} C : C \in \mathcal{C}\}$ is also locally finite. But if $C \in \mathcal{C}$, then there is a V_x in $\mathcal{V}$ with $C \subseteq V_x$. Thus, $\operatorname{cl} C \subseteq \operatorname{cl} V_x \subseteq G_x$, so that $\mathcal{F}$ is a locally finite closed cover that is a refinement of the original open cover $\mathcal{G}$. By the preceding lemma, X is paracompact. ∎

Proof of Michael's Theorem. For an open cover $\mathcal{G}$ of X let $\mathcal{A} = \bigcup_{n=1}^{\infty} \mathcal{A}_n$ be a cover of X, where each $\mathcal{A}_n$ is a locally finite collection of subsets of X and such that $\mathcal{A}$ is a refinement of $\mathcal{G}$. By the last lemma, to show that X is paracompact, we need only show that $\mathcal{A}$ has a locally finite refinement cover by not necessarily open sets.

For each $n \geq 1$ let

$$B_n = \bigcup \{A : A \in \mathcal{A}_n\}.$$

Put $B_0 = \emptyset$; since $\mathcal{A}$ is a cover, so is $\{B_n : n \geq 1\}$. Let

$$C_n = B_n \backslash B_{n-1}.$$

Note that: (1) $\{C_n : n \geq 1\}$ is a cover of X since $\{B_n : n \geq 1\}$ is; (2) the sets $C_1, C_2, \ldots$ are pairwise disjoint; (3) $C_n \subseteq B_n$. Let

$$\mathcal{D} = \{A \cap C_n : n \geq 1 \text{ and } A \in \mathcal{A}_n\}.$$

If $x \in X$, let n be the smallest integer such that there is a set A in $\mathcal{A}_n$ that contains x. Thus, n is the smallest integer such that $x \in B_n$, and it follows that $x \in C_n$. Hence, $\mathcal{D}$ is a cover of X. Clearly, $\mathcal{D}$ is a refinement of $\mathcal{A}$. Let $x \in X$, and again let n be the smallest integer such that there is a set A in $\mathcal{A}_n$ that contains x. Since $\mathcal{A}_n$ is locally finite, there is a neighborhood U of x that meets only a finite number of sets in $\mathcal{A}_n$. Since the sets $C_1, C_2, \ldots$

are pairwise disjoint, this says that U meets only a finite number of the sets in $\mathcal{D}$. That is, $\mathcal{D}$ is locally finite. By Lemma 3.7.8, X is paracompact.   ∎

Corollary 3.7.9. *If X is a regular topological space and $X = \bigcup_{n=1}^{\infty} X_n$ where each X_n is paracompact, then X is paracompact.*

Proof. If $\mathcal{G}$ is an open cover of X, then $\mathcal{G}_n = \{G \in \mathcal{G} : G \cap X_n \neq 0\}$ is an open cover of X_n. Since X_n is paracompact, there is an open cover $\mathcal{A}_n$ of X_n that is a refinement of $\mathcal{G}_n$. If $\mathcal{A} = \bigcup_{n=1}^{\infty} \mathcal{A}_n$, then $\mathcal{A}$ satisfies the condition in Michael's Theorem.   ∎

Corollary 3.7.10. *A separable metric space is paracompact.*

Proof. If (X, d) is separable and $\{a_1, a_2, \dots\}$ is a dense subset, let $\{r_1, r_2, \dots\}$ be an enumeration of the rational numbers in the open unit interval. So every open subset of X is the union of a collection of balls $\mathcal{B} = \{B(a_n; r_m) : n, m \geq 1\}$ (Corollary 2.2.4). Let $\mathcal{G}$ be an open cover of X. There is a countable subfamily $\mathcal{A}$ of $\mathcal{B}$ such that each open set G in $\mathcal{G}$ is the union of sets from some subcollection of $\mathcal{A}$. By letting $\mathcal{A} = \bigcup_{n=1}^{\infty} \mathcal{A}_n$, where each $\mathcal{A}_n$ consists of a single element, Michael's Theorem implies that (X, d) is paracompact.   ∎

It turns out that every metric space is paracompact. This is a theorem of Stone [12]. A simplified proof was found by Rudin [10]. Rudin's proof is indeed simple, but it uses the well-ordering principle, which I decided was something that in a course for this audience was not worth the exposition because we would not use it again. If you wish to learn about well ordering, then, with that under your belt, Rudin's paper is easily readable. The fact that separable metric spaces are paracompact (but with a different proof) was discovered by Dieudonné [3]. In fact, it was in this same paper that the concept of paracompactness was introduced and many properties were established.

Recall the definition of a σ-compact space given in the last section. Here is another consequence of Michael's Theorem.

Corollary 3.7.11. *If X is a regular σ-compact topological space, then X is paracompact.*

Theorem 3.7.12. *If X is locally compact, then X is paracompact if and only if $X = \bigcup\{X_i : i \in I\}$, where the sets $\{X_i\}$ are pairwise disjoint open σ-compact subsets.*

Proof. Assume $X = \bigcup\{X_i : i \in I\}$, as in the stated condition. Since X is locally compact, it is regular (Corollary 3.5.6). Thus, each X_i is paracompact by the preceding corollary. The fact that X is paracompact follows by Example 3.7.3(b).

Now assume X is paracompact. By first applying the fact that X is locally compact and then that it is paracompact, we have the existence of an open cover $\mathcal{U}$ of X that is locally finite and such that $\mathrm{cl}\, U$ is compact for each U in $\mathcal{U}$. (Verify!)

Claim. There is a subset C of X such that: (i) C is simultaneously open and closed; (ii) C is σ-compact.

We establish this claim by showing that there is a sequence of open subsets $\{G_n\}$ satisfying: (a) $\operatorname{cl} G_n \subseteq G_{n+1}$ for each $n \geq 1$; (b) each $\operatorname{cl} G_n$ is compact; (c) each G_n is the union of a finite number of sets from $\mathcal{U}$. This is done by induction. For G_1 take any set from $\mathcal{U}$. Now assume that we have $G_1, \ldots, G_n$ satisfying (a), (b), and (c). Since $\mathcal{U}$ is an open cover, there are sets $U_1, \ldots, U_m$ in $\mathcal{U}$ such that $\operatorname{cl} G_n \subseteq \bigcup_{k=1}^m U_k$; let $G_{n+1} = \bigcup_{k=1}^m U_k$. Clearly (a), (b), and (c) hold.

Now put $C = \bigcup_{n=1}^\infty G_n$. By the nature of the sets G_n, C is open and σ-compact. But since each G_n is the union of a finite number of sets from $\mathcal{U}$ and $\mathcal{U}$ is locally finite, we have that the collection $\{G_1, G_2, \ldots\}$ is locally finite. By Exercise 1, $\operatorname{cl} C = \bigcup_{n=1}^\infty \operatorname{cl} G_n \subseteq \bigcup_{n=1}^\infty G_n = C$, so C is also closed. This establishes the claim.

To finish the proof, we use Zorn's Lemma (Theorem A.4.6). Let $\mathcal{C}$ be the collection of all σ-compact subsets C that are simultaneously open and closed in X. By the claim, $\mathcal{C} \neq \emptyset$. Now let $\mathcal{W}$ be the collection of all subsets $\mathcal{S}$ of 2^X satisfying: (i) $\mathcal{S} \subseteq \mathcal{C}$; (ii) if $C_1, C_2 \in \mathcal{S}$ and $C_1 \cap C_2 \neq \emptyset$, then $C_1 = C_2$. Order $\mathcal{W}$ by inclusion. Since $\mathcal{C} \neq \emptyset$, $\mathcal{W} \neq \emptyset$. If $\mathcal{B}$ is a chain in $\mathcal{W}$, put $\mathcal{S}_0 = \bigcup\{\mathcal{S} : \mathcal{S} \in \mathcal{B}\}$. Clearly, $\mathcal{S}_0 \subseteq \mathcal{C}$ and, since $\mathcal{B}$ is a chain, condition (ii) of the definition of $\mathcal{W}$ is also satisfied by $\mathcal{S}_0$. Thus, Zorn's Lemma implies there is a maximal $\mathcal{S}_m$ in $\mathcal{W}$.

We want to show that $\bigcup\{C \in \mathcal{S}_m\} = X$. If this is not the case, then $\emptyset \neq Y = X \backslash \bigcup\{C \in \mathcal{S}_m\}$. Now Y is a closed subset of X and, therefore, is paracompact. Y is also an open set since $\mathcal{S}_m$ is a locally finite collection, so that $\operatorname{cl}(X \backslash Y) = \operatorname{cl} \bigcup\{C \in \mathcal{S}_m\} = \bigcup\{\operatorname{cl} C \in \mathcal{S}_m\} = \bigcup\{C \in \mathcal{S}_m\} = X \backslash Y$. According to the claim, there is a set C_0 in $\mathcal{C}$ such that $C_0 \subseteq Y$. Thus, $\mathcal{S}_m \cup \{C_0\} \in \mathcal{W}$ and is properly larger than $\mathcal{S}_m$, a contradiction. Therefore, $X = \bigcup\{C \in \mathcal{S}_m\}$, and the proof is complete. ∎

Example 3.7.13. The space $[0, \Omega)$ (§ 3.6) is not paracompact, though it is normal [Theorem 3.6.3(d)]. (See Exercise 4; the faint of heart can find this in [4], page 163, Example 3.)

Recall Theorem 3.3.10, where we proved that when we have a finite open cover of a normal space, we have a continuous partition of unity. Here we extend this to infinite covers, but we need to assume that the underlying space is paracompact. In a sense, it is the existence of partitions of unity as stated in Theorem 3.7.17 below that brings about our interest in paracompactness.

Definition 3.7.14. If X is a topological space, then a *partition of unity* is a collection of continuous functions $\{\phi_i : i \in I\}$ having the following properties.

(a) $\phi_i : X \to [0, 1]$ for all i in I.

(b)   The collection of sets $\{\{x \in X : \phi_i(x) > 0\} : i \in I\}$ is a locally finite open cover of X.

(c)   For every x in X, $\sum_i \phi_i(x) = 1$ (see below).

If $\mathcal{G}$ is an open cover of X, then we say that this partition of unity is *subordinate to the cover* $\mathcal{G}$ if the open cover $\{\{X \in X : \phi_i(x) > 0\} : i \in I\}$ is a refinement of $\mathcal{G}$.

We need to say a word about condition (c) in the definition. Note that for any x in X condition (b) implies that there is only a finite number of indices i in I such that $\phi_i(x) \neq 0$. So the sum in (c) is actually a finite sum. Even more dramatic, there is a neighborhood U of x such that $U \cap \{x \in X : \phi_i(x) > 0\} \neq \emptyset$ for only a finite number of i in I. This last, more dramatic, version will be useful when we apply partitions of unity.

How do we apply partitions of unity? That is a bit difficult to state in the abstract, but they are a means of putting together local results to obtain a global result. (Do not try to understand in detail that statement or what follows; just try to get the spirit of it.) For example, we might find that in a neighborhood of every point we can manufacture a continuous function with a certain property. The collection of all these neighborhoods constitutes an open cover of X. Then, assuming the underlying topological space X is paracompact and using Theorem 3.7.17 below, we manufacture a partition of unity $\{\phi_i\}$ subordinate to this cover. Multiplying each ϕ_i by the manufactured continuous function f_i for the neighborhood $\{x \in X : \phi_i(x) > 0\}$, we form $\sum_i \phi_i f_i$ to get a continuous function defined on all of X that has the property we have been seeking. (Again we use local finiteness to get that $\sum_i \phi_i f_i$ is actually a finite sum near each point and is therefore a continuous function.)

Before stating the theorem we seek, we state a pair of lemmas needed for its proof.

Lemma 3.7.15. *If X is paracompact and $\mathcal{U}$ is a locally finite open cover of X, then for every U in $\mathcal{U}$ there is an open set W_U such that* $\operatorname{cl} W_U \subseteq U$ *and* $\{W_U : U \in \mathcal{U}\}$ *is a locally finite open cover of X.*

Proof. Let $\mathcal{A}$ be the collection of all open sets A such that $A \cap U \neq \emptyset$ for only a finite number of the sets U in $\mathcal{U}$ and $\operatorname{cl} A$ is contained in at least one of these sets U. At this point it is not transparent that $\mathcal{A}$ is a nonempty collection. We will establish this and much more in what follows.

Claim. $\mathcal{A}$ is an open cover of X.

Fix an x in X. Since $\mathcal{U}$ is locally finite, there is a neighborhood G of x such that $\{U \in \mathcal{U} : x \in G \cap U\} = \{U_1, \ldots, U_n\}$. Thus, $G \cap \bigcap_{k=1}^n U_k$ is an open set that contains x. Since X must be normal (Proposition 3.7.5), there is an open set A with x in A and such that $\operatorname{cl} A \subseteq G \cap \bigcap_{k=1}^n U_k$. $A \in \mathcal{A}$, and this establishes the claim.

Now take a locally finite refinement $\mathcal{V}$ of $\mathcal{A}$. Temporarily fix a V in $\mathcal{V}$; since $\mathcal{V}$ is a refinement of $\mathcal{A}$, there is an A in $\mathcal{A}$ with $V \subseteq A$. By the definition of $\mathcal{A}$, there must be a set U in $\mathcal{U}$ such that $\operatorname{cl} V \subseteq U$. Thus, for every U in $\mathcal{U}$ we can define the nonempty open set

$$W_U = \bigcup \{V \in \mathcal{V} : \operatorname{cl} V \subseteq U\}.$$

Let $\mathcal{W} = \{W_U : U \in \mathcal{U}\}$. We must show that $\mathcal{W}$ is a locally finite open cover of X. First, from the definition of W_U we see that each V in $\mathcal{V}$ is contained in some W_U; since $\mathcal{V}$ is a cover, so is $\mathcal{W}$. Also, since $W_U \subseteq U$, $\mathcal{W}$ is a refinement of $\mathcal{U}$. To see that $\mathcal{W}$ is locally finite, let $x \in X$ and choose a neighborhood G of x that meets only a finite number of the sets U in $\mathcal{U}$. But if $G \cap W_U \neq \emptyset$, then $G \cap U \neq \emptyset$; hence G meets only a finite number of the sets in $\mathcal{W}$. Since x was arbitrary, $\mathcal{W}$ is locally finite. Because $\mathcal{W}$ is locally finite, Exercise 1 implies that $\operatorname{cl} W_U = \bigcup\{\operatorname{cl} V : V \in \mathcal{V} \text{ and } \operatorname{cl} V \subseteq U\} \subseteq U$. This completes the proof.   ∎

The next lemma has some independent interest.

Lemma 3.7.16. *If X is a normal topological space, $\mathcal{U}$ is a locally finite open cover of X, and $\mathcal{W} = \{W_U : U \in \mathcal{U}\}$ is a second open cover with the property that $\operatorname{cl} W_U \subseteq U$ for every U in $\mathcal{U}$, then there is a partition of unity on X subordinate to $\mathcal{U}$.*

Proof. For each U in $\mathcal{U}$ apply Urysohn's Lemma to obtain a continuous function $g_U : X \to [0,1]$ such that $g_U(x) = 1$ for every x in $\operatorname{cl} W_U$ and $g_U(x) = 0$ for x in $X \backslash U$. If $x \subset X$, choose a neighborhood G of x that meets only a finite number of sets in the cover $\mathcal{U}$, say $\{U_1, \ldots, U_n\} = \{U \in \mathcal{U} : U \cap G \neq \emptyset\}$. Thus, $\{g_{U_k} : 1 \leq k \leq n\}$ are the only functions in $\{g_U : U \in \mathcal{U}\}$ that do not vanish on G. Thus, $\sum_{U \in \mathcal{U}} g_U(y) = \sum_{k=1}^{n} g_{U_k}(y)$ is well defined and continuous on G. That is, each point x in X has a neighborhood such that on this neighborhood the function $g = \sum_{U \in \mathcal{U}} g_U$ is well defined and continuous. Moreover, since $x \in W_{U_k}$ for at least one k between 1 and n, we have that $g(x) \geq 1$ for each x in X. For each U in $\mathcal{U}$ let

$$\phi_U(x) = \frac{g_U(x)}{g(x)}.$$

It follows that ϕ_U is continuous. The reader can check that $\{\phi_U : U \in \mathcal{U}\}$ is a partition of unity subordinate to $\mathcal{U}$.   ∎

Theorem 3.7.17. *A topological space X is paracompact if and only if every open cover has a partition of unity subordinate to it.*

Proof. If $\mathcal{G}$ is an open cover and there is a partition of unity $\{\phi_i : i \in I\}$ subordinate to it, then $\{\{x : \phi_i(x) > 0\} : i \in I\}$ is a locally finite open cover that is a refinement of $\mathcal{G}$; hence X is paracompact.

The proof of the converse is a matter of putting together the preceding two lemmas with the fact that a paracompact space is normal. If $\mathcal{G}$ is an open cover, then let $\mathcal{U}$ be a locally finite refinement. There is a locally finite open cover $\mathcal{W} = \{W_U : U \in \mathcal{U}\}$ as in Lemma 3.7.15. Now use Lemma 3.7.16 to manufacture the partition of unity and check that it is subordinate to the original cover $\mathcal{G}$.   ∎

Proving in the metric space setting that every open cover has a partition of unity subordinate to it involves proving that the metric space is paracompact.

Many theorems prove that a topological space is metrizable. One that is easy to state is Smirnov's Theorem: X is metrizable if and only if it is paracompact and locally metrizable. (See Exercise 5.) This is a good example of the process referred to earlier in this section of using paracompactness to extend local results to a global result. We might also refer to Exercise 3 below, which gives a sufficient condition for a space to be paracompact. Along this line, the reader might also want to look at the references for the Nagata–Smirnov Metrization Theorem.

Exercises

(1) Show that if $\mathcal{S}$ is a locally finite collection of sets, then cl $[\bigcup\{S : S \in \mathcal{S}\}]$ $= \bigcup\{\operatorname{cl} S : S \in \mathcal{S}\}$. In particular, the union of a locally finite collection of closed sets is closed. (This extends Proposition 2.1.7(c) stated for finite unions.)

(2) Show that if $\mathcal{S}$ is a locally finite family of sets, then $\{\operatorname{cl} S : S \in \mathcal{S}\}$ is also locally finite.

(3) A topological space is called a *Lindelöf*[6] *space* if every open cover has a countable subcover. Prove that a regular Lindelöf space is paracompact.

(4) Verify Example 3.7.13.

(5) Say that a topological space is locally metrizable if every point has a neighborhood on which the relative topology is metrizable. Show that X is locally metrizable if and only if for each x in X and every neighborhood U of x there is a neighborhood G of x that is metrizable and contained in U.

---

[6]Ernst Leonard Lindelöf was born in 1870 at Helsingfors, Russian Empire, which is now Helsinki, Finland. (Helsinki had been under Swedish and Russian domination for a long time until Finland achieved independence in 1917 with the collapse of Russia.) His father was a professor of mathematics there, but by the time Lindelöf began to study at the university in 1887, his father was no longer on the faculty. In 1891 he went to study in Stockholm and in 1893–1894 in Paris. After this he returned to Helsingfors and graduated in 1895. In 1902 he became an Assistant Professor there, where, eventually as Professor, he remained until his retirement in 1938. The bulk of his research was in the theory of analytic functions, where several theorems of importance are due to him. His monograph *Le calcul des résidus et ses applications à la théorie des fonctions*, published in 1905, is still a valid reference that has been translated into several languages. In his later life he concentrated on teaching and writing several excellent texts. He died in 1946 in Helsinki.

Appendix

In this appendix we present some topics that arise during the course of the book. It is hard to tell what students starting the study of topology know outside of calculus, especially given the diverse paths that exist at a university. In fact, this is true in a given class at any given university, let alone students at different universities. Do all know the language of sets and functions? Do they know about the density of rational numbers? The list of questions could continue. So in this appendix we present some topics that start with sets and go up through a discussion of Zorn's Lemma, but mostly without proofs. I am fairly certain that most students who take this course will not have seen Zorn's Lemma, and I would advise all professors or instructors to cover this section—but not until it is needed in § 2.4. Also, don't look for a thorough treatment of the topics here. This appendix does not constitute a course in set theory or any other part of fundamental mathematics.

A.1. Sets

We want to talk a little about sets and functions in this section and the next so that we have a handy reference and to be sure that every reader has access to some of this material, which will be used without mention during the course of this book. Of course, this is not a complete exposition of this topic but merely a survey. We are going to assume some familiarity with the idea of a set and its subsets. In general we adopt what is often considered the naive approach to this topic, avoiding the rigorous logical approach.

We consider a set X and subsets $A, B, \ldots$ of X. For any subset A of X and any point x in X, the notation $x \in A$ means that x belongs to the set A. Similarly, the notation $x \notin A$ means that x is an element of X that does

J.B. Conway, *A Course in Point Set Topology*, Undergraduate Texts
in Mathematics, DOI 10.1007/978-3-319-02368-7,
© Springer International Publishing Switzerland 2014

not belong to A. We also talk of the *singleton* set, $\{x\}$, that is, the set that consists of the single point x. The *empty set* is the set consisting of nothing and is denoted by $\emptyset$. So there is no point x in X such that $x \in \emptyset$; equivalently, $x \notin \emptyset$ for every x in X.

If A and B are subsets of X, then the relation $A \subseteq B$ means that each x in A is also an element belonging to B; that is, if $x \in A$, then $x \in B$. Observe that $\emptyset \subseteq A$ and $A \subseteq X$ for any subset A of X. Similarly, $A \supseteq B$ means $B \subseteq A$. Note that if $A \subseteq B$ and $B \subseteq A$, then $A = B$. This gives a standard way of showing that two sets are equal. First take an element of A and show it belongs to B; that is, show that $A \subseteq B$. Then take an element of B and show it belongs to A; that is, show that $B \subseteq A$. This is put into action in the proof of Proposition A.1.3 below.

When $A, B \subseteq X$, we define the *union* of A and B as the set

$$A \cup B = \{x \in X : x \in A \text{ or } x \in B\}.$$

Note that the use of the word *or* here includes the possibility that x belongs to both A and B. In fact, if we ever need to exclude the possibility that x belongs to both sets, we will say this explicitly. There is in mathematics a term for the set that includes everything that belongs to A or to B but excludes those elements that belong to both; it's called the *symmetric difference*, but we will not use it.

Example A.1.1.   (a) If X is any set and $A \subseteq X$, then $A \cup \emptyset = A$ and $X \cup A = X$.

(b) If $X = \mathbb{R}$, the set of all real numbers, A is the open unit interval $(0, 1)$, and $B = (\frac{1}{2}, 3]$, then $A \cup B = (0, 3]$.

(c) If $X = \mathbb{R}^2$, the plane, A is the x-axis $\{(x, 0) : x \in \mathbb{R}\}$ and $B = \{(x, \frac{1}{2}) : x \in \mathbb{R}\}$, then $A \cup B = \{(x, y) : x \in \mathbb{R} \text{ and either } y = 0 \text{ or } y = \frac{1}{2}\}$.

The *intersection* of A and B is the set

$$A \cap B = \{x \in X : x \in A \text{ and } x \in B\}.$$

So $A \cap B \subseteq A \cup B$. When the sets A and B have no points in common, we say that the sets A and B are *disjoint* and write $A \cap B = \emptyset$.

Example A.1.2.   (a) If X is any set and $A \subseteq X$, then $A \cap X = A$ and $A \cap \emptyset = \emptyset$.

(b) If X, A, and B are as in Example A.1.1(b), then $A \cap B = (\frac{1}{2}, 1)$.

(c) If X, A, and B are as in Example A.1.1(c), then $A \cap B = \emptyset$.

Proposition A.1.3. *If X is any set and $A, B, C \subseteq X$, then:*

(a) $A \cap (B \cap C) = (A \cap B) \cap C;$

(c) $A \cup (B \cup C) = (A \cup B) \cup C;$

(c) $A \cap (B \cup C) = (A \cap B) \cup (A \cap C);$

(d) $A \cup (B \cap C) = (A \cup B) \cap (A \cup C).$

Proof. This is a good place to practice proving relations between sets. We will do part (c) in some detail, but proving the others is required of the reader in Exercise 1. The strategy here, which was alluded to earlier, is rather simple and straightforward. First we take an x on the left-hand side and show it belongs to the right-hand side, and then we do the opposite. This is the standard approach. Use it when an exercise asks you to show that two sets are equal.

(c) Let $x \in A \cap (B \cup C)$; so $x \in A$ and simultaneously $x \in B \cup C$. It follows that either $x \in B$ or $x \in C$. If it is the case that $x \in B$, then $x \in A \cap B$; if $x \in C$, then $x \in A \cap C$. That is, either $x \in A \cap B$ or $x \in A \cap C$, which is to say that $x \in (A \cap B) \cup (A \cap C)$.

Now assume that $x \in (A \cap B) \cup (A \cap C)$; so either $x \in A \cap B$ or $x \in A \cap C$. If $x \in A \cap B$, then $x \in A$ and $x \in B$; if $x \in A \cap C$, then $x \in A$ and $x \in C$. So under either alternative, $x \in A$; in addition, either $x \in B$ or $x \in C$. Thus, $x \subset A \cap (B \cup C)$. ∎

We also define the *difference* of the two sets as

$$A \backslash B = \{ x \in X : x \in A \text{ but } x \notin B \}.$$

Some mathematicians use the notation $A - B$ instead of $A \backslash B$. I prefer the notation $A \backslash B$ because in some situations $A - B$ is ambiguous. For example, if A and B are subsets of the real numbers, we will use the definition $A - B = \{ a - b : a \in A, b \in B \}$. The same applies when A and B are subsets of a vector space. So throughout this book the difference of two sets will be denoted using the backslash.

Example A.1.4. (a) If X is any set and $A \subseteq X$, then $A \backslash \emptyset = A$ and $A \backslash X = \emptyset$.

(b) If X, A, and B are as in Example A.1.1(b), then $A \backslash B = (0, \frac{1}{2}]$.

(c) If X, A, and B are as in Example A.1.1(c), then $A \backslash B = A$.

A notion that the reader may have encountered is the *complement* of a set: when $A \subseteq X$, the complement is $X \backslash A$. You can find various notations for this in the literature such as A^c or A' or $\tilde{A}$, but we will stick with $X \backslash A$. Of course, we have $X \backslash (X \backslash A) = A$, which some might interpret as two negatives make a positive.

Proposition A.1.5 (De Morgan's[1] Laws). *If X is any set and A and B are subsets of X, then:*

---

[1]Augustus De Morgan was born in 1806 in Madura, India (now Madurai). His father was an officer in the British army stationed there. The young Augustus lost his sight in one eye shortly after birth and the family returned to England when he was 7 months old. He entered Trinity College Cambridge in 1823. He received his bachelor's degree but refused to take a theology exam, which was required for an advanced degree. He returned to London in 1826 to study for the bar exam, but he became the first professor of mathematics at the newly established University College London despite the fact that he had never published in the subject. On a matter of principle he resigned in 1831. He was once again appointed in 1836, but he resigned again in 1866. In 1838 he introduced the term *mathematical induction*. The process had been in use before, but without clarity; De Morgan managed to give it a rigorous basis. Throughout his life he was a prolific writer with hundreds of articles and many books to his name. He introduced the laws of the present proposition and was a great reformer of mathematical logic. In addition, he founded the London Mathematical Society and became its first president. He

(a) $X\backslash(A \cup B) = (X\backslash A) \cap (X\backslash B)$;
(b) $X\backslash(A \cap B) = (X\backslash A) \cup (X\backslash B)$.

Proof. We prove (a) and leave the proof of (b) as Exercise 3. Again we use the standard approach to proving that two sets are equal. If $x \in X\backslash(A \cup B)$, then $x \notin A \cup B$. The only way this can happen is if both of the following two statements are true: $x \notin A$ and $x \notin B$. That is, $x \in X\backslash A$ and $x \in X\backslash B$; equivalently, $x \in (X\backslash A) \cap (X\backslash B)$.

Now assume that $x \in (X\backslash A) \cap (X\backslash B)$. This says that $x \in X\backslash A$ and $x \in X\backslash B$; that is, $x \notin A$ and $x \notin B$. But combining these two statements means $x \notin A \cup B$, or that $x \in X\backslash(A \cup B)$.   ∎

Finally, if X and Y are any sets, define the *cartesian product* of the sets as

$$X \times Y = \{(x, y) : x \in X, y \in Y\}.$$

In words, this is described as the set of all *ordered pairs* (x, y), where $x \in X$ and $y \in Y$. The use of the word *ordered* is meant to distinguish $X \times Y$ from $Y \times X$. In a similar fashion, if $X_1, \ldots, X_n$ are sets, then we define their cartesian product $X_1 \times \cdots \times X_n$ in a similar way.

In the next section we will define the cartesian product of an arbitrary collection of sets because to do that we need the idea of a function. But we can define now the union and intersection of an arbitrary collection of sets. Let X be a set, and suppose I is another set such that for each i in I we have a subset A_i of X; the set I is often called an *index set*, and we refer to the collection of subsets as $\{A_i : i \in I\}$. Define

$$\bigcup_{i \in I} A_i = \{x \in X : x \in A_i \text{ for at least one value of } i \text{ in } I\},,$$

$$\bigcap_{i \in I} A_i = \{x \in X : x \in A_i \text{ for all values of } i \text{ in } I\}.$$

Of course, when the index set I is the set of natural numbers $\mathbb{N} = \{1, 2, \ldots\}$, we use the notation

$$\bigcup_{n=1}^{\infty} A_n \quad \text{and} \quad \bigcap_{n=1}^{\infty} A_n.$$

The proof of the next result is left as Exercise 4.

Theorem A.1.6 (De Morgan's Laws). *If X is a set and $\{A_i : i \in I\}$ is a collection of subsets, then:*
(a) $X\backslash \left[\bigcup_{i \in I} A_i\right] = \bigcap_{i \in I}(X\backslash A_i)$;
(b) $X\backslash \left[\bigcap_{i \in I} A_i\right] = \bigcup_{i \in I}(X\backslash A_i)$.

was quite dogmatic, as his two resignations might indicate. He never voted and never visited the House of Commons, the Tower, or Westminster Abbey. He died in 1871 in London.

Exercises.

(1) Prove parts (a), (b), and (d) in Proposition A.1.3.
(2) Give a detailed proof that $X \backslash (X \backslash A) = A$.
(3) Prove part (b) of de Morgan's Laws.
(4) Prove Theorem A.1.6.

A.2. Functions

If X and Y are two sets, then a *function* from X into Y is a rule, denoted by $f : X \to Y$, that assigns to each x in X a unique point y in Y. Synonyms for function are the terms *map* and *mapping*. The set X is called the *domain* of f, and the set Y is called the *range* of f. The set $f(X) = \{f(x) \in Y : x \in X\}$ is called the *image* of f. Note the distinction between range and image. Now that you have noted the distinction, you should be aware that some mathematicians define the range of a function as what we are calling the image and vice versa. When they do this, they sometimes use the term *codomain* for what we call the range. Confused? Don't worry too much about it except when you consult other sources; we will consistently use the terms as we defined them here.

Example A.2.1. (a) $f : \mathbb{R} \to \mathbb{R}$ defined by $f(x) = x^2$ is a function. Its domain and range are $\mathbb{R}$, and its image is $[0, \infty)$.
(b) If for each x in $\mathbb{R}$ we let $f(x) = +1$ when $x \geq 0$ and $f(x) = -1$ when $x \leq 0$, then this is not a function since the value of $f(0)$ is not uniquely defined. If we were to redefine f by stating that $f(x) = -1$ when $x < 0$, then it would be a function.
(c) If X and Y are sets, $y_0 \in Y$, and $f(x) = y_0$ for every x in X, then $f : X \to Y$ is a function —called a constant function.
(d) If X is a set and $A \subseteq X$, define $\chi_A : X \to \mathbb{R}$ by

$$\chi_A(x) = \begin{cases} 1 & \text{if } x \in, A \\ 0 & \text{if } x \notin A. \end{cases}$$

This function is called the *characteristic function* of A. Some call this the *indicator function*. Observe that for all x in X, $\chi_\emptyset(x) = 0$ and $\chi_X(x) = 1$.

The set of all functions from X into Y is denoted by

$$Y^X.$$

Consistent with this is the notation 2^X used to denote the set of all subsets of X. This consistency results from identifying the collection of all subsets with the set of all characteristic functions from X into the two-point set $\{0, 1\}$.

If $f : X \to Y$ and $g : Y \to Z$, then the *composition* of f and g is the function $g \circ f : X \to Z$ defined by

$$g \circ f(x) = g(f(x))$$

for all x in X. So, for example, if $f(x) = x^2$ and $g(x) = \sin x$, then $g \circ f(x) = \sin(x^2)$, while $f \circ g(x) = (\sin x)^2$.

Definition A.2.2. A function $f : X \to Y$ is *injective* or *one-to-one* if whenever $x_1, x_2 \in X$ and $f(x_1) = f(x_2)$, then $x_1 = x_2$. The function is said to be *surjective* or *onto* if for every y in Y there is an x in X such that $f(x) = y$. A function is *bijective* if it is both injective and surjective.

In this book we will use the terms injective and surjective. There is a linguistic objection to using "onto" here as the term "an onto function" is bad English; onto is a preposition, not an adjective. Some people are passionate about avoiding onto; I would not say I am passionate about this, but I do care. Hence I use surjective. I use injective to maintain the symmetry of the terms.

Example A.2.3. (a) If $\pi : \mathbb{R}^2 \to \mathbb{R}$ is defined by $\pi(x, y) = x$, then π is surjective but not injective.

(b) If X is any set and A is a nonempty subset of X, then the *inclusion map* $i : A \to X$ is defined by $i(a) = a$ for all a in A. This map is always injective but fails to be surjective unless $A = X$, in which case it is called the *identity function*.

(c) The map $f : \mathbb{R} \to (-\pi/2, \pi/2)$ defined by $f(t) = \arctan t$ is bijective.

(d) The map $f : \mathbb{R} \to \mathbb{R}$ defined by $f(t) = \sin t$ is neither injective nor surjective. If we consider the sine function as mapping $\mathbb{R}$ into $[-1, 1]$, then it is surjective; if we consider it as a mapping from $[-\pi/2, \pi/2]$ into $\mathbb{R}$, it is injective but not surjective; if we consider it as a mapping from $[-\pi/2, \pi/2]$ to $[-1, 1]$, the sine function is bijective. So by changing the domain or the range or both of a function, we can often have it acquire one of the properties of the last definition.

If $f : X \to Y$ is a bijective function, then we can define the *inverse* of f as follows. If $y \in Y$, then by surjectivity there is an x in X with $f(x) = y$; by injectivity, this x is unique. Thus, we can define $f^{-1}(y) = x$ for the unique point x with $f(x) = y$. This makes $f^{-1} : Y \to X$ into a function.

The proof of the next proposition is Exercise 2.

Proposition A.2.4. *Let $f : X \to Y$ and $g : Y \to Z$ be functions.*

(a)   *If f is bijective, then $f \circ f^{-1}$ is the identity function on Y and $f^{-1} \circ f$ is the identity function on X.*

(b)   *If both f and g are bijective, then so is $g \circ f$ and $(g \circ f)^{-1} = f^{-1} \circ g^{-1}$.*

We conclude this section by using the concept of a function to define the cartesian product of an arbitrary collection of sets.

Definition A.2.5. If $\{X_i : i \in I\}$ is a collection of sets, then the *cartesian product* or just *product* of the sets is defined by

$$\prod_{i \in I} X_i = \left\{ x : I \to \bigcup_{i \in I} X_i \text{ such that for all } i, x(i) \in X_i \right\}.$$

First note that this is consistent with the definition of the product of a finite number of sets given in the preceding section. Indeed, if X_1 and X_2 are two sets and we define $X_1 \times X_2$ as in § A.1, then for each (x_1, x_2) in $X_1 \times X_2$ we can define a function $x : \{1, 2\} \to X_1 \cup X_2$ by $x(1) = x_1, x(2) = x_2$ and vice versa. So even though the idea of an ordered pair has been suppressed, it is implicit in the use of $\{1, 2\}$ as the domain of these functions. We will have more to say on arbitrary products in § 2.6.

Exercises.

(1) Consider the composition of the two functions $g \circ f$, look at the nine possible situations where f and g each have one of the properties injective, surjective, and bijective, and ascertain which, if any, of these properties is possessed by $g \circ f$.
(2) Prove Proposition A.2.4.

A.3. The Real Numbers

Here we collect a few facts about the real numbers, some of which readers might know and some which they might not. It may also be that the reader has a subliminal understanding about some properties of the real numbers, and this section will help bring those into a more vivid focus. In addition, the real numbers not only serve as an inspiration for the study of many topological spaces, but their properties help us to discover many important results about continuous functions from a topological space into the space of real numbers. So my advice is to give this section at least a quick read.

The set of real numbers, $\mathbb{R}$, has its usual order $\leq$. If $E \subseteq \mathbb{R}$, then we say that E is *bounded above* if there is a number a such that $x \leq a$ for all x in E. Such a number a is called an *upper bound* of E. Similarly, E is *bounded below* if there is a number b with $b \leq x$ for all x in E; b is called a *lower bound* of E. It is easy to see that E is bounded above by an upper bound a if and only if the set $-E = \{-x : x \in E\}$ is bounded below by a lower bound $-a$. For this reason any statement about the upper bound of a set has its analog for the lower bound, and we will frequently only do the upper version. A set E is *bounded* if it is bounded both above and below.

Definition A.3.1. If $E \subseteq \mathbb{R}$ and E is bounded above, then a *least upper bound* or *supremum* of E is a number α that satisfies the following conditions:

(i) α is an upper bound for E;
(ii) $\alpha \leq a$ for any other upper bound a for E.

Similarly, if E is bounded below, then the *greatest lower bound* or *infimum* of E is a number β that satisfies the following conditions:

(i) β is a lower bound for E;
(ii) $b \leq \beta$ for any other lower bound b for E.

In symbols we write $\alpha = \sup E$ and $\beta = \inf E$. (The reader may have seen the notation $\alpha = \mathrm{lub}\ E$ and $\beta = \mathrm{glb}\ E$, but we will use the sup and inf notation.)

It is straightforward that the supremum or infimum of a set, if it exists, is always unique (Exercise 1).

An interesting project is to give a rigorous development or definition of the real numbers starting with the basic axioms of the positive integers $\mathbb{N}$. The most common way to do this is to define the rational numbers $\mathbb{Q}$ using standard algebra and then introduce the concept of Dedekind cuts for the rational numbers. We will not carry this out here but instead will just assume that we are given the real numbers with all their properties. Two fundamental properties are as follows, which we take as axioms but would follow from a rigorous definition and development.

Axiom A.3.2 (Density Property). *If a and b are rational numbers and $a < b$, then there is an irrational number x with $a < x < b$. Similarly, if a, b are irrational numbers and $a < b$, then there is a rational number x with $a < x < b$.*

Axiom A.3.3 (Completeness Property). *If a nonempty subset E of $\mathbb{R}$ has an upper bound, then it has a supremum.*

These may seem obvious to you, but that is probably because you have always thought of $\mathbb{R}$ as having these properties. Unless, however, you are in possession of an exact definition of the real numbers such as the set of all Dedekind cuts, you cannot give a rigorous proof of the Completeness Property. Let us also remark that the set $\mathbb{Q}$ does not have the Completeness Property. For example, $\{q \in \mathbb{Q} : q^2 < 2\}$ does not have a supremum within the set $\mathbb{Q}$. Of course, it has a supremum in $\mathbb{R}$, namely $\sqrt{2}$.

It follows from the Completeness Property that if a nonempty subset E of $\mathbb{R}$ is bounded below, then $\inf E$ exists. In fact, as mentioned earlier, this follows by applying Axiom A.3.3 to the set $-E$.

The next proposition, in spite of its trivial proof, will be most useful and applied frequently.

Proposition A.3.4. *If $E \subseteq \mathbb{R}$ has a supremum α and $\epsilon > 0$, then there is an x in E with $\alpha - \epsilon < x \le \alpha$. Similarly, if $\beta = \inf E$, then there is a y in E with $\beta \le y < \beta + \epsilon$.*

Proof. In fact, by the definition of supremum, $\alpha - \epsilon$ cannot be an upper bound for E, so the existence of x follows. The statement about the infimum can be proven by applying the first part of the set $-E$. See Exercise 1. ∎

Be aware that when we say a sequence $\{x_n\}$ is *increasing*, we mean that $x_n \le x_{n+1}$ for all n. Some call this a *nondecreasing* sequence, but in spite of the accuracy, I have always found that term counterintuitive, or at least cumbersome. In this terminology, a constant sequence is increasing. We will use the term *strictly increasing* to denote a sequence $\{x_n\}$ that satisfies $x_n < x_{n+1}$ for all n. A sequence $\{x_n\}$ is *decreasing* or *strictly decreasing* if the sequence $\{-x_n\}$ is, respectively, increasing or strictly increasing.

Corollary A.3.5. *If E is a set that is bounded above and $\alpha = \sup E$, then there is an increasing sequence $\{x_n\}$ in E such that $x_n \to \alpha$. Similarly, if E is a set that is bounded below and $\beta = \inf E$, then there is a decreasing sequence $\{y_n\}$ in E such that $y_n \to \beta$.*

Proof. We will establish the following claim.

Claim. For each $n \geq 1$ there is an x_n in E such that $x_1 \leq \cdots \leq x_n$ and $\alpha - n^{-1} < x_n \leq \alpha$.

We establish this by induction. We get x_1 by applying the proposition with $\epsilon = 1$. Assume we have $x_1, \ldots, x_n$ with the stated properties. If one of the x_k equals α, choose $x_{n+1} = \alpha$. Assume that $x_k < \alpha$ for $1 \leq k \leq n$, and let $0 < \epsilon < \min\{(n+1)^{-1}, \alpha - x_1, \ldots, \alpha - x_n\}$. By the proposition , we can find a point x_{n+1} in E with $\alpha - \epsilon < x_{n+1} \leq \alpha$. It follows that x_{n+1} has the properties listed in the claim when n is replaced by $n + 1$.

Once we have the claim, we can apply the definition of convergence for a sequence to conclude the proof. The statement about the infimum can be proven by applying the first part to the set $-E$. See Exercise 1. ∎

A fact that emerges from close examination of the proof of the preceding corollary is that if $\alpha \notin E$, then the sequence $\{x_n\}$ can be chosen to be strictly increasing.

In one sense, the next result is the converse of the last corollary, but strictly speaking it is not.

Proposition A.3.6. *If a bounded sequence is either increasing or decreasing, then it must converge.*

Proof. Let us assume that $\{x_n\}$ is an increasing sequence in $\mathbb{R}$ that is bounded. (Note that in this case this amounts to assuming the sequence is bounded above since x_1 is automatically a lower bound for the sequence.) By the Completeness Property, $\alpha = \sup\{x_1, x_2, \ldots\}$ exists. If $\epsilon > 0$, then the preceding proposition says there is an integer N with $\alpha - \epsilon < x_N \leq \alpha$. Since the sequence is increasing, we have that $\alpha - \epsilon < x_n \leq \alpha$ whenever $n \geq N$. Thus, $|x_n - \alpha| < \epsilon$ for all $n \geq N$, so that $x_n \to \alpha$. The proof when the sequence is decreasing is left to the reader. ∎

Exercises.

(1) (a) If E is a subset of $\mathbb{R}$ that is bounded above, show that the supremum is unique. (b) If $\alpha = \sup E$, show that $-\alpha = \inf\{-E\}$.
(2) In Corollary A.3.5, show that if $\alpha \notin E$, then the sequence $\{x_n\}$ can be chosen to be strictly increasing.

A.4. Zorn's Lemma

This is a section I believe will be unfamiliar to almost every reader. It is included in the appendix only because it is not a part of topology, though it is essential to know in order to prove some of the most important results in that subject. It is essential to master this material.

Definition A.4.1. A partially ordered set is a pair $(S, \leq)$, where S is a set and $\leq$ is a relation on the elements of S that has the following properties for all x, y, z in S: (i) $x \leq x$ (*reflexivity*); (ii) if $x \leq y$ and $y \leq x$, then $x = y$ (*antisymmetry*); (iii) if $x \leq y$ and $y \leq z$, then $x \leq z$ (*transitivity*).

Example A.4.2.   (a) The real line is an example of a partially ordered set.
  (b) If $(S, \leq)$ is a partially ordered set and $T \subseteq S$, then $(T, \leq)$ is a partially ordered set.
  (c) If X is any set and 2^X is the collection of all subsets of X, define $\leq$ on 2^X to be containment. That is, for A, B in 2^X, $A \leq B$ means that $A \subseteq B$. This makes $(2^X, \leq)$ a partially ordered set.
  (d) From (b) and (c) we see that any collection $\mathcal{G}$ of subsets of X is also a partially ordered set. For example, if $(X, \mathcal{T})$ is a topological space and $\mathcal{G}$ is the collection of all open sets, then $(\mathcal{G}, \subseteq)$ is a partially ordered set.
  (e) Let X be a set, and let $\mathbb{R}^X$ denote the collection of all functions $f : X \to \mathbb{R}$. If $f, g \in \mathbb{R}^X$, define $f \leq g$ to mean that $f(x) \leq g(x)$ for all x in X. With this definition $(\mathbb{R}^X, \leq)$ is a partially ordered set.
  (f) Suppose $(S, \leq)$ is a partially ordered set, and define a new relation $\lesssim$ on S by saying that $x \lesssim y$ means $y \leq x$. It follows that $(S, \lesssim)$ is a partially ordered set.

As can be expected, when we have a partially ordered set $(S, \leq)$ and $x, y \in S$, the notation $y \geq x$ means $x \leq y$.

Definition A.4.3. If $(S, \leq)$ is a partially ordered set and $T \subseteq S$, an *upper bound* for T is an element x in S (but not necessarily in T itself) such that $x \geq y$ for every y in T. A *maximal element* for T is an element x of T such that T contains no larger element. That is, x is a maximal element if whenever $y \in T$ and $x \leq y$, then $y = x$.

Note that when x is a maximal element of T, it is possible for there to be an upper bound y of T that is different from x, but if this happens, we have that $y \notin T$. That is, $y \in S \backslash T$.

Definition A.4.4. A partially ordered set $(S, \leq)$ is called *linearly ordered* if for any elements x and y in S either $x \leq y$ or $x \geq y$. Some mathematicians say that a linearly ordered set is *simply, completely,* or *totally* ordered.

Of course, the real line is a linearly ordered set. Also, see Exercise 2.

Definition A.4.5. If $(S, \leq)$ is a partially ordered set, then a *chain* in S is a subset that is linearly ordered.

After this long string of definitions, we are now in a position to state Zorn's Lemma.

Theorem A.4.6 (Zorn's[2] Lemma). *If $(S, \leq)$ is a partially ordered set such that every chain in S has an upper bound, then S has a maximal element.*

We will not prove this; the reader is asked to accept it as a given truth. The proof takes us too far from the philosophy and objectives of this book. Actually, the proof relies on the Axiom of Choice, which most mathematicians accept as a given fact of set theory. In fact, in its simplest formulation, this axiom seems obviously true. It states that if you are given an arbitrary collection of nonempty sets $\{A_i : i \in I\}$, it is possible to choose one element a_i from each set. Nevertheless, it is known that the Axiom of Choice does not follow as a consequence of the standard axioms of set theory. So we assume it is true. A proof that Zorn's Lemma and the Axiom of Choice are equivalent is available in Theorem 3.12 of [5]

In what follows, we will give an application of Zorn's Lemma, but first an example where it does not apply. Consider the partially ordered set $\mathbb{R}^X$ defined in Example A.4.2(e). If we take $\mathcal{C} = \{f_1, f_2, \dots\}$, where each f_n is the constant function $f_n(x) = n$, then $\mathcal{C}$ is a chain, but it has no upper bound. Therefore, Zorn's Lemma does not apply. Here is an application of the use of Zorn's Lemma.

Consider an arbitrary vector space $\mathcal{X}$ over the real numbers. Recall that a (possibly infinite) set of vectors B in $\mathcal{X}$ is linearly independent if for any finite subset $\{x_1, \dots, x_n\}$ of B the only scalars $a_1, \dots, a_n$ that satisfy $\sum_{k=1}^n a_k x_k = 0$ are $a_1 = \dots = a_n = 0$. If $\mathcal{M}$ is the collection of all linearly independent sets of vectors, we use set inclusion to make $\mathcal{M}$ into a partially ordered set as in Example A.4.2(c). A *basis* for $\mathcal{X}$ is a maximal linearly independent set of vectors.

Once you have a basis for $\mathcal{X}$, it follows that every vector in $\mathcal{X}$ can be written as a unique linear combination of basis elements (Exercise 4). Proving

---

[2]Max August Zorn was born in Krefeld, Germany, near Dusseldorf, in 1906. He received his Ph.D. from Hamburg University in 1930, having worked under Emil Artin. He was forced to leave his first position at Halle in 1933 because of Nazi anti-Semitic policies. He emigrated to the USA with an appointment at Yale University, where he proved his famous lemma. Later he made contributions in analysis with an investigation of the differentiability of functions defined on a Banach space. In 1936 he went to the University of California at Los Angeles, where he remained until 1946; he then went to Indiana University, where I had the pleasure of being his colleague for many years. We lived a block apart and occasionally walked home together. He retired in 1971. Like some, he retired from teaching and committee work, but not from mathematics. Until his death he remained involved with mathematics, not publishing but attending a long list of seminars and every colloquium in the department. He had the practice of always asking a question at the end of a lecture. My experience when I gave my interview talk at Indiana was typical. The question he asked me seemed weird initially, though I gave something like an answer. Thirty minutes later what he was asking finally hit me; it was insightful and worth thinking about. Needless to say, I went to see him about it. I am not sure all his questions were relevant and insightful, but as far as I could see, most were. Nevertheless, the speakers, especially those from outside Indiana University, frequently had the same befuddled reaction I did. In 1993 he died after being hit by an automobile while crossing the street in front of the mathematics department.

the existence of a basis in the finite-dimensional setting is not difficult, but in the infinite-dimensional case it requires Zorn's Lemma.

Theorem A.4.7. *Every vector space over* $\mathbb{R}$ *has a basis.*

Proof. Let $\mathcal{M}$ be the collection of all linearly independent subsets of $\mathcal{X}$ and order $\mathcal{M}$ by inclusion: if $A, B \in \mathcal{M}$, then $A \leq B$ means that $A \subseteq B$. Let $\mathcal{C}$ be a chain in $\mathcal{M}$ and put $B = \bigcup\{A : A \in \mathcal{C}\}$. Let $x_1, \ldots, x_n \in B$; so there are $A_1, \ldots, A_n$ in $\mathcal{C}$ such that $x_k \in A_k$ for $1 \leq k \leq n$. Since $\mathcal{C}$ is a chain one of the sets contains all the others (Exercise 3); denote this biggest set by A_k. Thus, for $1 \leq j \leq n$, $x_j \in A_k$, and so $\{x_1, \ldots, x_n\}$ is linearly independent. That is, $B \in \mathcal{M}$. Clearly, B is an upper bound for the chain, so that Zorn's Lemma applies and proves the existence of a maximal linearly independent set. ∎

The preceding theorem is phrased for vector spaces over $\mathbb{R}$, but the same concepts and proof work for a vector space over any field.

Exercises.

(1) Verify that each of the examples in Example A.4.2 is a partially ordered set.
(2) Which of the partially ordered sets in Example A.4.2 are linearly ordered?
(3) If $(S, \leq)$ is a linearly ordered set and A is any finite subset of S, then there is an element a in A with $x \leq a$ for each x in A.
(4) Prove that if $\mathcal{X}$ is a vector space, then a set B of linearly independent vectors in $\mathcal{X}$ is a basis (a maximal linearly independent set) if and only if for each x in $\mathcal{X}$ there are unique vectors $x_1, \ldots, x_n$ in B and unique scalars $a_1, \ldots, a_n$ such that $x = \sum_{k=1}^{n} a_k x_k$.
(5) (This exercise requires that you know the definition of a group.) If G is a group and H is an abelian subgroup, then there is a maximal abelian subgroup K of G such that $H \subseteq K$.

A.5. Countable Sets

In this section we will explore the notion of a countably infinite set, the smallest of the orders of infinity. The fact that not all infinite sets are equivalent is one that comes as a surprise to many people. Indeed, historically this was a shock to the world of mathematics when Cantor[3] first revealed it.

Definition A.5.1. A set X is *countable* if there exists a subset A of the natural numbers $\mathbb{N}$ and a bijective function $f : A \to X$.

Example A.5.2. (a) Any finite set is countable. For infinite sets that are countable, we will say they are *countably infinite*. Some say that such an infinite set is *denumerable*. Below we show the existence of sets that are not countable.

---

[3]See the biographical note to Theorem 1.2.12.

(b) The set of all integers, $\mathbb{Z}$, is countable. In fact, $0, 1, -1, 2, -2, 3, -3, \ldots$ describes a bijective function from $\mathbb{N}$ onto $\mathbb{Z}$. For convenience we will often show a set is countable by describing how to exhaust the set by writing it as a sequence, as we just did. Such an undertaking tells us how to define a bijective function even though finding a formula for that function may be far from clear. In the present case, it is not difficult to write a formula for this function. Indeed if $n \in \mathbb{N}$ and we set

$$f(n) = \begin{cases} \frac{n}{2} & \text{if } n \text{ is even,} \\ -\frac{n-1}{2} & \text{if } n \text{ is odd,} \end{cases}$$

then $f : \mathbb{N} \to \mathbb{Z}$ is a function that gives the correspondence described earlier. In other situations, writing a formula for the function may range from challenging to impossible. The point of this discussion is that proving the existence of such a function does not mean we have to write a formula for the function. If we describe a process or algorithm for determining which element of a set corresponds to each integer and if this process exhausts the set, then we have described the required function.

(c) Any subset of a countable set is countable. In fact, this is immediate from the definition of a countable set.

The next proposition is useful in showing that a given set is countable.

Proposition A.5.3. (a) *If X is any set such that there is a subset A of $\mathbb{N}$ and a surjective function $f : A \to X$, then X is countable.*

(b) *If X is a countable set, Y is a another set, and there is a surjection $f : X \to Y$, then Y is countable.*

Proof. To prove (a), let f and A be as in the statement. For each x in X let n_x be the first integer n in A with $f(n) = x$; that is, $n_x = \min f^{-1}(x)$. Thus, $B = \{n_x : x \in X\}$ is another subset of $\mathbb{N}$ and $g : B \to X$ defined by $g(n_x) = x$ is a bijection.

Part (b) is immediate from (a) and the definition of a countable set. ∎

The next result will be quite helpful in proving that some sets are countable. See, for example, the subsequent corollary.

Proposition A.5.4. *If X is countable, then so is $X \times X$.*

Proof. If X is finite, then the result is easy. Thus, assume that X is infinite. To prove the proposition , it is equivalent to show that $\mathbb{N} \times \mathbb{N}$ is countable. (Why?) Here we want to define a bijection $f : \mathbb{N} \to \mathbb{N} \times \mathbb{N}$. Again we need only show how to arrange the elements (m, n) in $\mathbb{N} \times \mathbb{N}$ in a sequence. So imagine $\mathbb{N} \times \mathbb{N}$ as an infinite square array of pairs of positive integers. In the first row are all the pairs $\{(1, n) : n \in \mathbb{N}\}$; in the second $\{(2, n) : n \in \mathbb{N}\}$; and so forth. We write down the following sequence of entries:

$$(1, 1), (2, 1), (1, 2), (3, 1), (2, 2), (1, 3), (4, 1), (3, 2), (2, 3), (1, 4), \ldots.$$

If you write the array on paper and draw northeast diagonal lines connecting these pairs, you should be able to discern the pattern. (Many other patterns are possible.) This describes the bijection.   ∎

Corollary A.5.5. *The set of rational numbers is countable.*

Proof. Writing each rational number as a fraction in reduced terms we see that there is a bijection between $\mathbb{Q}$ and a subset of $\mathbb{Z} \times \mathbb{Z}$, which is countable by the proposition.   ∎

Proposition A.5.6. *If $X = \bigcup_{n=1}^{\infty} X_n$ and each of the sets X_n is countable, then X is countable.*

Proof. We write $X_n = \{x_n^1, x_n^2, \dots\}$; if X_n is infinite, we can do this with $x_n^k \neq x_n^j$ for all $k, j \geq 1$. Otherwise, repeat one of the points an infinite number of times. Thus, $f : \mathbb{N} \times \mathbb{N} \to X$ defined by $f(n, k) = x_n^k$ is surjective. It follows by Proposition A.5.3(b) that X is countable.   ∎

Corollary A.5.7. *The set of all finite subsets of $\mathbb{N}$ is countable.*

Proof. If F denotes the set of all finite subsets of $\mathbb{N}$, then note that $F = \bigcup_{n=1}^{\infty} S_n$, where S_n is the set of all subsets of $\{1, 2, \dots, n\}$. But S_n is a finite set. In fact, from combinatorics we know that S_n has 2^n elements. By the preceding proposition, F is countable.   ∎

Now we turn to some results showing the existence of uncountable sets.

Proposition A.5.8. *The set of all sequences of zeros and ones is not countable.*

Proof. Let X be the set of all sequences of zeros and ones, and suppose it is countable; thus, we can write $X = \{x_1, x_2, \dots\}$. We manufacture an element a in X such that $a \neq x_n$ for any $n \geq 1$. This will furnish a contradiction to the assumption that we have an exhaustive list and thus proves the proposition. Suppose that for each $n \geq 1$, $x_n = x_n^1 x_n^2 \cdots$ is a sequence of zeros and ones; in other words, $x_n = \{x_n^k : k = 1, 2, \dots\}$. If $n \geq 1$ and $x_n^n = 0$, let $a^n = 1$; otherwise, let $a^n = 0$. This defines an $a = a^1 a^2 \dots$ in X. Since $a^n \neq x_n^n$, $a \neq x_n$ for any $n \geq 1$. This gives our desired contradiction.   ∎

Corollary A.5.9. *The collection of all subsets of $\mathbb{N}$, $2^{\mathbb{N}}$, is not countable.*

Proof. In fact, by looking at the characteristic functions of subsets of $\mathbb{N}$, we see that the set of all sequences of zeros and ones is in bijective correspondence with $2^{\mathbb{N}}$.   ∎

To prove the next proposition, we must consider the dyadic expansions of numbers in the unit interval, a topic we also use in § 3.3. For $0 \leq x \leq 1$ we can write

$$x = \sum_{n=1}^{\infty} \frac{x_n}{2^n},$$

where each $x_n = 0$ or 1. (This series always converges since $\sum_{n=1}^{\infty} 2^{-n} = 1$.)
The proof that each x in the unit interval can be so expanded is not too
complicated and proceeds as follows. Consider x, and divide the interval into
its equal halves: $[0, \frac{1}{2}]$ and $[\frac{1}{2}, 1]$. If x belongs to the first half, let $x_1 = 0$;
if $x \in [\frac{1}{2}, 1]$, let $x_1 = 1$. (We note an ambiguity here if $x = \frac{1}{2}$, and we will
address this shortly.) Note that in either case we have that $|x - x_1/2| < \frac{1}{2}$.
Now consider whichever half interval contains x and divide it into two equal
halves; let $x_2 = 0$ if x belongs to the first half and $x_2 = 1$ if it belongs to the
second half. Now we have that

$$\left| x - \left(\frac{x_1}{2} + \frac{x_2}{2^2} \right) \right| < \frac{1}{2^2}.$$

Continue this process, and we see that the series so defined will converge to
x. (The reader who wants to write out the details can formulate an induction
statement based on what we just did and prove it. See Exercise 3.)

What about the ambiguity? If $x = a/2^n$ for some $n \geq 1$ and $0 < a < 2^n$,
then the choice of x_n can be either 0 or 1. This is the only way such an
ambiguity arises since, in fact, using the summation for a geometric series,

$$\sum_{k=n}^{\infty} \frac{1}{2^k} = \frac{1}{2^n} \sum_{k=0}^{\infty} \frac{1}{2^k} = \frac{1}{2^n} \frac{1}{1 - \frac{1}{2}} = \frac{1}{2^{n-1}}.$$

It follows that if $\{x_n\}, \{y_n\}$ are two sequences of zeros and ones, then the
only way we can have that $\sum_{n=1}^{\infty} x_n/2^n = \sum_{n=1}^{\infty} y_n/2^n$ is for one sequence
to end in all zeros and the other to end in all ones. See Exercise 4.

Proposition A.5.10. *The interval $(0, 1)$ is not countable.*

Proof. In a sense, this proposition is a corollary of Proposition A.5.8, but its
proof is a bit more involved than you usually associate with a corollary. Let
X be the set of all sequences of zeros and ones that are not constantly zero
from some point on. We observe that there is an injective mapping between
X and the open interval $(0, 1)$ by the discussion preceding the statement
of the proposition. On the other hand, by considering the characteristic
functions of subsets of $\mathbb{N}$, there is a bijective mapping between this set of
sequences and the infinite subsets of $\mathbb{N}$. By Corollaries A.5.9 and A.5.7, X is
not countable. ∎

Exercises.

(1) Show that if A is an infinite subset of $\mathbb{N}$, X is a countably infinite set,
and $f : A \to X$ is a bijection, then there is a bijection $g : \mathbb{N} \to X$. (Hint:
first show that if A is an infinite subset of $\mathbb{N}$, then there is a bijection
$h : \mathbb{N} \to A$.)

(2) Show that if $X_1, \ldots, X_n$ are countable sets, then so is $X_1 \times \cdots \times X_n$.

(3) Write out a detailed proof that each x in the unit interval has a dyadic
expansion.

(4) If $\{x_n\}, \{y_n\}$ are two sequences of zeros and ones, show that $\sum_{n=1}^{\infty} x_n/2^n$ $= \sum_{n=1}^{\infty} y_n/2^n$ if and only if there is an integer n such that $x_k = 0$ and $y_k = 1$ for all $k \geq n$.

(5) Show that if $X_1, X_2, \ldots$ are countably infinite sets, then $\prod_{n=1}^{\infty} X_n$ is not countable.

References

[1] Conway, J.B.: The inadequacy of sequences. Am. Math. Monthly **76**, 68–69 (1969)

[2] Conway, J.B.: A Course in Abstract Analysis. American Mathematical Society, Providence (2012)

[3] Dieudonné, J.: Une généralisation des espaces compacts. J. Math. Pures Appl. **23**, 65–76 (1944)

[4] Dugundji, J.: Topology. Allyn and Bacon, Newton (1967)

[5] Hewitt, E., Stromberg, K.: Real and Abstract Analysis. Springer, New York (1975)

[6] Kelley, J.L.: General Topology. Ishi, New York (2008)

[7] Michael, E.: A note on paracompact spaces. Proc. Am. Math. Soc. **4**, 831–838 (1953)

[8] Michael, E.: Another note on paracompact spaces. Proc. Am. Math. Soc. **8**, 822–828 (1957)

[9] Raha, A.B.: An example of a regular space that is not completely regular. Proc. Math. Sci. **102**, 49–51 (1992)

[10] Rudin, M.E.: A new proof that metric spaces are paracompact. Proc. Am. Math. Soc. **20**, 603 (1969)

[11] Steen, L.A., Seebach, J.A.: Counterexamples in Topology. Springer, New York (1978)

[12] Stone, A.H.: Paracompactness and product spaces. Bull. Am. Math. Soc. **54**, 977–982 (1948)

J.B. Conway, *A Course in Point Set Topology*, Undergraduate Texts in Mathematics, DOI 10.1007/978-3-319-02368-7,
© Springer International Publishing Switzerland 2014

Terms

J.B. Conway, *A Course in Point Set Topology*, Undergraduate Texts
in Mathematics, DOI 10.1007/978-3-319-02368-7,
© Springer International Publishing Switzerland 2014

Symbols

J.B. Conway, *A Course in Point Set Topology*, Undergraduate Texts
in Mathematics, DOI 10.1007/978-3-319-02368-7,
© Springer International Publishing Switzerland 2014

Printed in the United States
By Bookmasters